CW01202761

Neuromarketing in Action

Neuromarketing in Action

How to talk and sell to the brain

Patrick M Georges,
Anne-Sophie Bayle-Tourtoulou
and Michel Badoc

KoganPage

LONDON PHILADELPHIA NEW DELHI

Publisher's note

Every possible effort has been made to ensure that the information contained in this book is accurate at the time of going to press, and the publishers and authors cannot accept responsibility for any errors or omissions, however caused. No responsibility for loss or damage occasioned to any person acting, or refraining from action, as a result of the material in this publication can be accepted by the editor, the publisher or any of the authors.

First published in Great Britain and the United States in 2014 by Kogan Page Limited

Apart from any fair dealing for the purposes of research or private study, or criticism or review, as permitted under the Copyright, Designs and Patents Act 1988, this publication may only be reproduced, stored or transmitted, in any form or by any means, with the prior permission in writing of the publishers, or in the case of reprographic reproduction in accordance with the terms and licences issued by the CLA. Enquiries concerning reproduction outside these terms should be sent to the publishers at the undermentioned addresses:

2nd Floor, 45 Gee Street	1518 Walnut Street, Suite 1100	4737/23 Ansari Road
London EC1V 3RS	Philadelphia PA 19102	Daryaganj
United Kingdom	USA	New Delhi 110002
www.koganpage.com		India

© Patrick M Georges, Anne-Sophie Bayle-Tourtoulou and Michel Badoc, 2014

The right of Patrick M Georges, Anne-Sophie Bayle-Tourtoulou and Michel Badoc to be identified as the authors of this work has been asserted by them in accordance with the Copyright, Designs and Patents Act 1988.

ISBN   978 0 7494 6927 6
E-ISBN   978 0 7494 6928 3

British Library Cataloguing-in-Publication Data

A CIP record for this book is available from the British Library.

Library of Congress Cataloging-in-Publication Data

Bayle-Tourtoulou, Anne-Sophie.
   Neuromarketing in action : how to talk and sell to the brain / Anne-Sophie Bayle-Tourtoulou, Patrick M. Georges and Michel Badoc.
      pages cm
   ISBN 978-0-7494-6927-6 (pbk.) – ISBN 978-0-7494-6928-3 (ebk.)   1. Neuromarketing.
I. Georges, Patrick, 1953- II. Badoc, Michel. III. Title.
   HF5415.12615.B39 2013
   658.8001'9–dc23
                                                                                    2013028705

Typeset by Graphicraft Limited, Hong Kong
Printed and bound in India by Replika Press Pvt Ltd

CONTENTS

ACKNOWLEDGEMENTS

We would like to thank all the managers, staff, professors, physicians, ergonomists, knowledge engineers, researchers and so on who worked with the authors.

We also want to thank the Paris Chamber of Commerce and Industry and HEC Paris, as well as the companies that collaborated with us, in particular Unilever, Accor, Promod and SAP.

Finally our thanks go to Dominique Rouziès (Professor at HEC Paris), Laurent Maruani (Marketing Department Chair, HEC Paris), Michael Segalla (Professor at HEC Paris), Didier Reynaud (Affiliance) and Philippe Honoré (Taly) for their help and encouragement, as well as Jean-François Guillon (Chief Operating Officer, HEC Executive Château), who has helped us a lot in the diffusion of Neuromarketing to European firms.

Introduction

In the United States, the first extensive Neuromarketing study was launched between 2004 and 2007. It involved 2,081 volunteers in five countries: the United States, the UK, Germany, Japan and the Republic of China. These volunteers agreed to magnetic resonance imaging (MRI) to see how their brain responded to different marketing stimuli.

The initiator, Martin Lindstrom, an international expert in branding policy, was fascinated by the application of techniques derived from neuroscience to marketing. In his book reflecting the results of his experiments, *Buyology* (2010) (which since then, despite criticism regarding its methodology, has become a best-seller), Lindstrom points out certain shortcomings of traditional marketing studies. He reflects on the fact that, despite the tens of billions of dollars spent on market research in the United States, the failure rate of new products amounts to nearly 80 per cent three months after the launch. He wonders why, despite the hundreds of billions of dollars invested in advertising at a growing rate during the past 40 years, the advertisement recall rates continue to decline. Recall scores for television can now be below 5 per cent only minutes after the advertisement appeared. The allocated budgets are however increasingly substantial.

These are some of the reasons why this author started searching for more efficient marketing techniques than the traditional studies used thus far. Neuromarketing was beginning to take off. Its purpose is to use the advances made by neuroscience to improve the understanding of consumer behaviour, in particular the consumers' emotions, which hold the key to a large part of their purchasing decisions, but also their memorization and their positive or negative desires. This set of processes can be detected in a more reliable manner when one examines how the brain works, or how it gives its instructions to the human body by secreting hormones or 'neurotransmitters'.

In addition to this important research, the 'industrialization' of Neuromarketing has developed throughout the world. Certain companies, such as NeuroFocus, acquired by Nielsen, or NeuroInsight in the United States, Neurosense Ltd in the UK, SalesBrain in the United States and France, NeuroInsight in Australia, Brain Impact in Belgium and many others, are developing Neuromarketing research tools.

Some of these companies claim they can propose more predictive marketing techniques than traditional market studies. In Laurence Serfaty's very interesting film *Des citoyens sous influence* (Citizens under the influence), which can be seen on YouTube, AK Pradeep, NeuroFocus's director, claims he is able to anticipate the success or failure of a communication campaign based on his Neuromarketing investigations. Stanford's Brian Knutson, in his laboratory in Palo Alto, can predict, by looking at how the brain lights up on the MRI, a decision to purchase before the subject expresses it.

We are progressively entering the Neuromarketing era. Its purpose is not to replace but to complete the traditional studies used thus far in this discipline. By appealing directly to the brain, Neuromarketing avoids the biases of the questionnaire, the interlocutor, and the difficulty in describing emotions or what our senses express. This bias is particularly encountered when interviewing people on sensitive subjects such as racism, money or sex. Unlike spoken statements, the brain does not lie. The lit-up part actually corresponds with the emotion felt.

Marketing is above all the science of perception more than reality or logic. In this context, as pointed out by Martin Lindstrom, a large part of our purchasing is essentially linked to our emotions.

This trend is accentuated when dealing with young populations, often referred to as generation X or Y. It appears that emotion is progressively replacing logic, while channel surfing is replacing reasoning. The spectacular development of the internet and the virtual and interactive world reinforces this trend every day.

It could be useful to trace the source of emotions and decisions by improving the understanding of how the brain and human intelligence work. This is the purpose of Neuromarketing, which is destined to enhance today's marketing and play an increasingly important role in tomorrow's marketing.

About the authors

Patrick M Georges is a professor in management at the Collège des Ingénieurs in Paris. He is also a medical doctor and neurosurgeon, and former head of the neurosurgery department of the University Medical Centre Vésale in Belgium. He specializes in cognitive and decision sciences. He created the management cockpit, the favourite decision support system with managers, distributed worldwide by the giant software maker SAP. He is a management books author at Business Expert Press, New York. He is the scientific director of the company NEURO, a scientific network that screens neuroscience research publications to provide marketing managers with advance leads and clues for new products, messages and sales approaches. His website is **www.patrick-georges.net**.

Professor Anne-Sophie Bayle-Tourtoulou has taught marketing within the different programmes of HEC Paris for many years. She is the Academic Director of the HEC Marketing Major and the HEC Master of Sciences in Marketing. Since her doctoral thesis, she has specialized in the retail domain. She works on specific marketing characteristics in this sector and focuses on issues relating to assortment, retailers' own brands, pricing and promotional policies, and stock shortages. As a consultant for SymphonyIRI, she has also developed expertise in panels and published managerial studies and academic articles on the above-mentioned subjects. In light of the confirmed and rapid development of the internet and the new issues and challenges this poses for professionals and scholars, she explores the changes retailers have made to revive and reinvent links between their sales areas and consumers, or between their personnel, key actors of customer relationships, and consumers. The benefit of neuroscience for the sensory marketing of sales outlets and customer relationship management opens up promising and relatively unexplored fields of research.

Professor Michel Badoc has been teaching marketing for several years, mostly at the various institutions of the HEC Group, but also at universities, business schools and vocational schools such as CESB or ENASS. He develops appraisal and consultancy activities for European and North American companies. Having worked for several years as a consultant for TNS Sofres, he is fascinated by all that makes it possible to better understand consumers' – including socially conscious web consumers' – behaviour. He takes a special interest in clients' emotional, instinctive and irrational purchasing habits. He presents to his students a set of methods aimed at improving the relevance of their oral and written presentations to directors and executive committees. He studies new forms of communication that address the saturation phenomenon affecting advertising messages. Lastly, he takes an interest in the application to marketing of new theories that boost innovation. He is the author of several publications.

For what readership is the work intended?

A Deloitte study (Lee and Stewart, 2012) mentioned Neuromarketing as one of the 12 emerging trends that will mark technologies, the media and telecommunications. Since the publication of this study, this new marketing discipline is on the way to arousing interest on a worldwide scale. Neuromarketing consists in applying techniques stemming from neurosciences (MRI, electroencephalography, hormone secretion, telemetry, etc) to marketing.

Neuromarketing in Action is a fundamental work for readers wishing to learn more about this recent marketing advance. As said before, it is the fruit of close collaboration between Patrick M Georges, a neurosurgeon practising in Belgium, and Anne-Sophie Bayle-Tourtoulou and Michel Badoc, two marketing professors working in France. The three authors teach at HEC, a top European business school that was ranked first internationally in 2013 for the training of top-level executives (HEC Executive Management).

After explaining how the brain functions and the ways in which it unconsciously influences consumers' behaviour, the book describes possible applications in the various marketing fields of today (eg studies, product alteration, selling methods, sensory marketing at the point of sale, communications) and tomorrow (eg value innovation, creation of sensory brands, improved interactivity with social networks, permission marketing). It features several examples of practical applications in Europe and across the world.

The book is intended for a wide readership: for people working in marketing, surveys, communications, sales and distribution, and for the many students in these fields and also in psychology, neurosciences, neuromedicine and so on. It is also meant for consumer protection bodies, public authorities, political parties and, more generally, citizens wishing to make smart purchases by gaining better knowledge of the way in which their brain reacts to the numerous enticements used in marketing, sales and communications.

PART I
Neuromarketing or the art of selling to the brain

As a science studying how to bring companies closer to their customers, marketing has numerous limitations in terms of studies as well as business approach, sales and communication. By drawing on neuroscience, which helps probe human intelligence and comprehends the unconscious of the brain, it significantly improves its efficacy with all its interlocutors: managers, employees, partners and, of course, customers.

This new approach constitutes the Neuromarketing domain. It was born out of the technical possibilities, inspired by the medical sector, of analysing how the brain works and their applications to marketing. Studies in neuromedicine largely rely on the possibility of lighting up the lobes of the brain associated with decision making and action. They were identified by a wealth of medical research relating to epilepsy, Parkinson's disease and other forms of brain injury. Studies also draw on the analysis of hormone secretion and its impact on human behaviour. For instance when dopamine is largely secreted, it provides pleasure and makes the desire to buy products greater. Neuromarketing studies help in understanding how the brain responds to different stimuli and making decisions.

The use of neuroscience in the marketing approach, because of its efficiency, is not without danger. It can be used properly only by marketers with sound ethics and irreproachable professional conduct.

Marketing and its limitations in understanding human intelligence

01

Marketing represents an analytical tool, a state of mind, an approach, and technical expertise. As with a sporting event that one watches while comfortably seated in the stadium or in front of the television, its practice may, at first glance, seem simple, if not simplistic. This is a misleading illusion. Success is the result of patented professionalism as well as serious predispositions. Like high-level athletes, marketing professionals achieve success by learning good technical approaches and by developing their practice. The speed of performance linked to the switch from amateurism to professionalism largely depends on the acquisition of good marketing gestures. To acquire these gestures, marketing professionals must hone their skills in this discipline, and master the methods and tools, while being aware of their limitations. Neuroscience can enable them to push back these limitations to improve their efficiency. Its contribution to the marketing discipline leads to the emergence of a new discipline: Neuromarketing.

The concepts of marketing and Neuromarketing

Marketing to acquire a sense of the customer

The influence of marketing requires acknowledging that manufacturing products or services are no longer ends in themselves but means to the end of satisfying customer needs. This new form of relation-ship with customers leads to a radical change of attitude.

From a marketing perspective, the consumer's tastes and needs prevail over those of the technicians. Customer needs are becoming the number one source of inspiration in the development of products and services.

While marketing is above all a state of mind designed to resolutely focus all company resources on customer satisfaction, the company must not realize this desire to the detriment of its own best interests, in particular the two critical imperatives, ie profitability and main-taining or improving its image. This is why we limit ourselves to defining marketing as:

THE ART OF SATISFYING ONE'S CUSTOMERS WHILE MAKING THE BOSS HAPPY (profitability–quality–image)

or

THE ART OF CREATING VALUE BOTH FOR THE CUSTOMERS AND FOR THE COMPANY

The obligation to adapt to the consumers' tastes, needs and expecta-tions is even more crucial when the company performs its activities in an economic environment where the offer is greater than the demand, an environment often referred to as a 'market economy'.

Increasing competition gives consumers a greater choice and makes them more demanding. It is sometimes said that the customer is king. The development of new information and communication techno-logies (NICTs) based on the internet only accentuates this trend. Possible choices and information prior to a purchase have increased exponentially. The opportunities are such that consumers can even switch from the status of king to that of dictator. With the internet, they are no longer satisfied with one-way information and com-munication. They want to ask questions and share their views. They switch progressively from the status of consumer to that of proactive consumer. They are guided by conscious as well as subconscious expectations, emanating from their brain's responses to different stimuli. Thorough knowledge of neuroscience helps marketing pro-fessionals comprehend the subconscious motives that help consumers decide and act. *The integration of the two disciplines opens the paths of Neuromarketing.*

The Neuromarketing space

The role of marketing is not to decide but to provide insight for managers and operational staff so that they can make the right decisions. It must act as a spotlight for decision makers, who often have to make do with a torch to guide their decision. The benefit of neuroscience is that, by adapting this light to how the brain works, it helps transform lighting into conviction.

The art of marketing is to give decision makers in headquarters and in the field a better understanding of their environment in order to make the right choices. It consists of finding space to enable the creation of value for both the customer and the company. Neuromarketing integrates the study of conscious and subconscious motivations, which can lead to a decision, into this space (see Figure 1.1). Motivations are apparent in customers as well as business decision makers. To succeed, Neuromarketing must be capable of convincing customers, so that they buy, and decision makers, so that they agree to give marketing the budget and resources necessary to please, convince and attract customers.

FIGURE 1.1   The Neuromarketing space

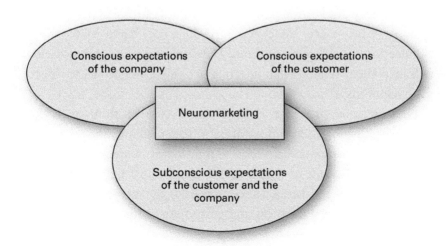

Internal decision makers must be convinced first. Without their support, the marketer will not have the means required to win over the customer. Moreover, the marketing function must often teach the internal organization how to be customer-centric and approach the customer in the best possible conditions.

The approach that makes it possible to find this space is difficult to implement, as it requires confronting these two partners with conflicting and sometimes contradictory requirements. To achieve this, genuine professionalism based on a rigorous approach is crucial.

To be effective, professionalism relies on a body of knowledge derived from marketing and neuroscience that sheds light on the conscious and subconscious modes of action of customers and internal decision makers.

Evolution within the Neuromarketing space calls for both modesty and scientific rigour. It helps ensure that managers do not extrapolate their subjectivity to the entire market. They begin to adopt a marketing attitude when they no longer consider themselves a representative sample of their customer base, regardless of their experience. Marketing professionals focus on listening to the customer rather than holding preconceived or subjective ideas. Thorough knowledge of how the brain comprehends and processes the information received from the environment helps them improve their objectivity.

Rigorous and methodical marketing practitioners must also have a salesperson's temperament. They must be able to listen and understand their interlocutors and make themselves understood by adapting the way they think and speak to their interlocutors' communication patterns. This is true in particular when they have to work with engineers, financial personnel and IT people. The use of neuroscience will help them realize what can instinctively guide these partners in their reactions.

Neuroscience provides an interesting contribution, as it gives additional insight, sometimes different from declarations. This knowledge helps distinguish between the interlocutors' deep thoughts and conventional attitude or even doublespeak. It sheds light on the brain's stimuli that trigger, often subconsciously, the positive or negative reactions of internal partners and customers to the decision.

Marketing limitations and the contribution of neuroscience: the path of Neuromarketing

The marketing issue can be illustrated by the humorous albeit pertinent cartoon presented in Figure 1.2. The role of the marketing director or marketing staff (product, market, brand managers) is to implement a policy designed to remove both '?'s: that of the elephant, which symbolizes the customer, and that of the employee.

FIGURE 1.2 Marketing issue

SOURCE: McDonald and Morris (1992).

Remove the elephant's '?'

To do this, the primary role of the marketing function is to provide the company with insight into what customers expect from their suppliers: knowing what goes through their mind, and their tastes, needs and expectations. This is necessary to avoid any errors when designing the different elements of the 'marketing mix', for example development and presentation of the product or service, pricing, sale, distribution, communication and after-sale. The marketing function must, in this respect, be a guiding light providing further, broader, more detailed insight into what customers think. Most of the time, managers have at first only a narrow view of these needs, largely conditioned by their preconceived ideas. They then have a battery of quantitative and qualitative consumer studies at their disposal.

Unfortunately, these studies have numerous limitations. Based on statements, they reflect only what the customers say, which is not necessarily what they think. For difficult topics such as politics, racism, sexuality or other delicate subjects, customers sometimes conceal their actual thoughts for fear of offending their interlocutor or being criticized for having ideas that do not conform with the common view. In other cases, they are incapable of comprehending certain key elements that influence their purchase, for example the colours of a yogurt pot, the shape of the packaging, or the label. These oversights, which may seem insignificant to professionals, sometimes lead the marketing function to make errors in the assessment of most marketing mix elements. In this context, thorough knowledge of how the brain reacts to the different stimuli in its environment can provide an additional vision, making it possible to adapt the companies' offers and propositions to the consumers' actual, unexpressed expectations. The evaluation of instinctive behavioural reactions improves the explanation of certain purchasing trends that may seem irrational in the customers' decision-making process. The use of neuroscience as part of Neuromarketing helps complete the results from traditional studies and provide marketers with a more in-depth vision of the needs felt by customers.

Looking into what happens in the brain significantly improves the marketing knowledge relating to the consumer's behaviour, purchasing processes and communication perception. The pertinence of Neuromarketing studies is particularly obvious when researchers focus on the emotion triggered in customers, their recall, their positive or negative desires, and their predispositions to purchase.

Remove the employee's '?'

While it provides insight, marketing is seldom a source of decisions. While its role is to guide the different decision makers on what customers expect and what they do or do not like, its role is rarely to decide on their behalf, unless the marketing director is also the company's CEO or managing director, which is often the case in small and medium-sized enterprises but not in large corporations.

To gain support for a shared market vision, the marketing manager needs to convince. This need to convince is all the more important as the customer's expectations often relate to perception more than reality or rationality. In a world where employees are rational (engineers, physicians, IT staff, experts from various departments, etc), it can be difficult to show that, even when customers are wrong, perception is what ultimately drives them to purchase a product or service. It is difficult to get competent and rational individuals to understand that certain details they deem insignificant, such as the colour or shape of packaging, the location on the shelf, the telephone greeting or the smile of a person behind a counter, can affect sales by 20 to 50 per cent.

For an identical offer, the marketing manager must convince the principal decision makers, generally the members of the company's executive committee, of the importance of all the elements guiding the consumers in their decisions. This manager must find persuasive arguments and present them in a form that allows every member of the committee to retain the essence of these arguments. Extensive knowledge of the human brain's memory storage modes and of the elements likely to impregnate or influence the brain is a key success factor for any convincing presentation. Managers must also be able to efficiently manage project teams. They can use the tools derived from neuroscience to their advantage, with a view to developing a marketing-specific 'management cockpit' and facilitating the decision-making process. They must convince colleagues from other departments and, of course, salespeople of the relevance of the ideas designed to satisfy customers. When a decision is made to adapt the company to the customers' expectations, managers must organize and drive the changes necessary to adjust the employees' attitudes while limiting stress.

Using the variety of persuasion tools derived from neuroscience can help managers with this task. The purpose of the following recommendations is not to replace the traditional presentation methods already used by the marketing function, such as neurolinguistic programming (NLP), but rather to complete them by reflecting on what aetiological knowledge, especially applied to the brain, can bring to this discipline. This knowledge, which also has its limitations, does not claim to provide an absolute vision, but sheds a different light, which can be combined with other approaches. Neuromarketing, a cross between marketing and neuroscience, is designed to provide this unique and complementary insight.

Cognitive ergonomics to help the marketing function answer its questions

Cognitive ergonomics is the science of people at work. It seeks to adapt work to people with a view to improving their productivity and satisfaction. In marketing, it is used to adapt the product to the customer and therefore improve customer satisfaction. The product is more likely to please and be purchased if it is adapted to human perception, human vision, human memory and the brain's decision-making mechanisms.

Experts in cognitive ergonomics are increasingly familiar with what it takes for the brain to accept a product as 'good'. Of course, marketing departments have been designing best-selling products for a long time without the help of neuroscientists. The marketing function has been using ergonomics unknowingly for decades. The only difference is that experts in cognitive ergonomics have created an actual method to make ergonomic design or correction faster, more systematic, more efficient and less costly.

This method does not aim at replacing those traditionally used by marketing but rather at completing them. It will get more and more important in marketing research as knowledge about how the brain and human intelligence work grows rapidly and as its application in marketing becomes a reality in international firms.

Neuroscience as a way to discover the secrets of human intelligence

02

Neuroscience, also known as 'cognitive science', studies how the brain works and how it can change. It gives us crucial information on how it responds to stimuli or inhibitions. For the past few years, it has provided the foundation of a body of research and publications listed by Professor Patrick M Georges and his colleagues as part of the NET Research programme in Belgium. Its application to business management is the subject of several books (Georges, 2004b; Lecerf-Thomas, 2009). Its application to marketing and sales is more limited (Renvoisé and Morin, 2002).

As we do not wish to give an exhaustive description of how the brain traditionally operates, we will limit ourselves to a few key ideas. Readers interested in exploring this issue further will find in the References a number of publications to enhance their knowledge. Certain ideas inspired by neuroscience sometimes relate to 'popular wisdom'. Our objective is to legitimize certain practices, derived from 'popular wisdom', based on scientific verifications rendered possible by new methods of investigation, for example by lighting up the brain or analysing hormone secretions.

Studies and tools inspired by neuroscience

Advances in medicine, computer sciences and radiology pave the way for new studies. The stimulation and lighting up of certain parts of the brain via an MRI or electrodes, the analysis of secreted hormones, and the use of micro-sensors in telemetry make it possible to gain a better understanding of how the brain responds to various stimuli, as well as to measure the stress and stress tolerance levels of employees or customers. These studies, which directly emanate from neuroscience, were reserved for medical and therapeutic use for a long time. They have only recently emerged in the marketing domain for understanding the behaviour of companies' employees as well as customers. Certain consumer goods and services companies, while remaining quiet on this issue, are beginning to use these studies in addition to traditional market surveys.

The brain is a complex organism. The studies conducted in an attempt to enhance our understanding of the brain are still in the early stages. Fundamental discoveries are bound to emerge in the 21st century that will profoundly modify knowledge relating to consumer behaviour and psychology and will shed light on certain automatic or impulsive reactions that are still unexplained. Knowledge of these reactions leads to significant progress in medical treatments. This knowledge is also used by the police and the judiciary to understand certain reactions in the event of high stress levels such as in assault or rape. Marketing, which in essence is a science focusing on knowledge of customers and proactive customers, must develop its understanding of these studies.

'Image studies' to read the customer's mind

How do we know what the customer really thinks? When a neuron works, it produces electricity, needs more blood to nourish itself and its metabolism changes. These are three things that modern Neuromarketing devices can detect. We can therefore obtain images of the brain areas at work when we perform a given action, when we have a given thought or when we make a given decision.

How do we know, for instance, that the function of a given area is music? The principle is fairly simple: if, in all patients who have lost the notion of music as a result of an accident, a 'hole' is always observed at the same place on the images of their brain, it is reasonable to assume that this area is in charge of music.

However, these areas are not always specific: they can light up for various reasons that are not always clearly understood by scientists.

The principal brain visualization tools

The most commonly used brain visualization tools include:

- *Functional magnetic resonance imaging (fMRI):* The principle of this non-traumatic technique is to detect the changes in blood oxygenation that occur in response to neuronal activity. It provides beautiful pictures of the brain and helps accurately locate the cerebral areas activated during a cognitive, behavioural or emotional task.

- *Electroencephalography (EEG):* This technique measures, with great timing accuracy, the electric field of the brain by means of electrodes placed on to the scalp.

- *Hyper-scanning, ie an internet evolution of the fMRI* (Le Bihan, 2012): This makes it possible to get two people to play together and assess what their mental strategy is when responding to one another, even if these people are at opposite ends of the Earth. The latest marketing experiments in hyper-scanning show the live reactions of the brains of a customer and a salesperson talking to each other. The salesperson, by looking into the customer's brain, can adapt the offer not only to what the customer is telling him or her but also to what the customer is thinking. If the salesperson realizes that using a specific argument lights up the customer's dislike region, he or she instantly adapts the speech used, even if the customer has not remarked upon it out of politeness. While it is of course impossible to perform an MRI on all customers, these experiments teach us valuable lessons, such as when and how to announce the price during a product presentation.

What lights up during an fMRI

The human brain (see Figure 2.1) is divided into areas, which have each been given scientific names. If the brain were a cube, the frontal lobe would be the front surface, the parietal lobe the rear surface, the frontal lobes the lateral surfaces and the limbic lobe the lower surface.

FIGURE 2.1   Vision of the brain

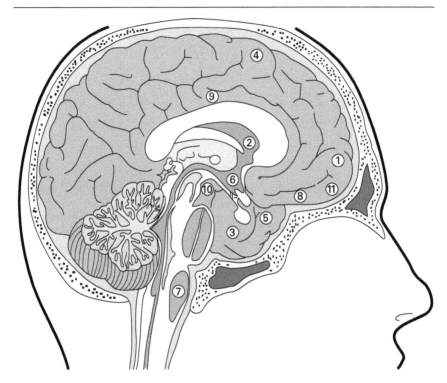

KEY: 1. Prefrontal cortex: memory, deduction, planning. 2. Hypothalamus: happiness. 3. Limbic cortex: emotions, surprise. 4. Premotor cortex: imitation. 5. Temporal cortex: memory, empathy. 6. Brain amygdala: aggressiveness. 7. Endorphin neurons: pain. 8. Ventromedial prefrontal cortex: behaviours. 9. Cingulate cortex: envy, sadness. 10. Hippocampus: stress. 11. Orbital cortex: emotional control.

Neuroscience has progressively revealed the functions of the different areas of the brain which, being highly complex because of its systemic nature, continues to be studied on a daily basis.

During an fMRI, areas of the brain 'light up' according to the tasks or stimuli proposed. Thus:

- The nucleus accumbens, the brain's pleasure centre, lights up when a person is shown what he or she really desires, something precious to the person: alcohol, sex, a game, food, or related products that can herald this pleasure. It lights up for fatty and sweet food, sexual pleasure and beautiful cars (a sign of wealth and therefore of the ability to protect the offspring).

- The lateral prefrontal cortex lights up when we are asked to decide, when will prevails over instincts.

- The amygdalae are activated when something scares us or makes us aggressive or when we are anxious.
- The limbic cortex kicks in when we are moved emotionally and are remembering.
- The premotor area is activated when we see a movement and prepare to imitate it.
- The occipital cortex is activated when we look.
- The temporal cortex is activated when we listen.
- The ventral putamen is activated when we experience a feeling of satisfaction.
- The medial prefrontal cortex is activated when we assess a value.

Medical imaging studies are fairly complex. The equipment is leased from hospitals or private clinics during periods when it is not required for medical purposes, for the one or two days required by the marketing study. This involves a significant cost: around €10,000 per day, which includes the engineer. Similarly, purchasing an MRI machine, as certain major corporations have done for the purposes of Neuromarketing studies, is expensive: approximately €2 million. Another element to be taken into account is that, during an MRI, the customer is lying in a tube in a hospital. This is a far cry from the purchasing context as a whole, and this limitation must be kept in mind. The electroencephalogram does not encounter this problem, as the experiment can be conducted directly in the sales outlet, with the electrodes hidden under a hat. However, as this technique is extremely sensitive to surrounding movements or those of the person wearing the electrodes, it can sometimes be unreliable within a natural as opposed to a laboratory context.

Analysis of hormones, moods or secretions to gain insight into the customer's response to marketing stimuli

When confronted with stimuli, the brain produces an increasing level of hormones, which can be measured in our body fluids: urine, saliva or blood. There is a variety of hormones linked to pleasure, aggressiveness, stress, serenity, etc. Below is a non-exhaustive list of hormones that can be measured almost continuously in a customer about to make a decision:

- Serotonin, the good mood hormone, protects against depression and impulsiveness.
- Dopamine facilitates impulsiveness and aggressiveness. It is also a sign of pleasure.
- Cortisol in saliva measures stress intensity.
- Testosterone is linked to sexual desire.
- Progesterone and oestrogen are extremely important in women.
- Noradrenalin creates excitement and shared pleasure.
- Adrenalin triggers tension and stress.
- Endorphins condition well-being and self-composure.

Therefore every desire and feeling of pleasure is correlated with three hormones or neurotransmitters: dopamine (tension of desire), noradrenalin (excitement, shared pleasure) and endorphins (well-being, rest).

If you want to taste serotonin, all you have to do is put a broad smile on your face, mouth taut and eyes screwed up. Your emotional or limbic brain, uninformed by your voluntary brain, which produced the facial movement, immediately reacts: 'Oh, he/she is smiling! I must have been distracted and missed a happy event. Quick, I need an injection of serotonin to adjust this!' And you will experience joy, a fleeting but very 'real' feeling. Try it and you will see that it works two or three times, and then less and less. Your limbic brain, albeit instinctive, is not stupid and understands that you are playing with it.

Measuring hormones is an accurate and inexpensive process. However, measurements of blood or urine samples are not very practical, which is less the case for saliva measurements.

The sectors resorting to these hormone secretion studies include the culinary industry. Everyone knows that a menu that makes you salivate helps sell dishes. Marketing knows how to find the words that sell dishes, notably by using saliva tests: by putting a salivation detector in the customer's mouth (which is easy, cheap and comfortable) and testing various dish names based on the amount of saliva generated, the preferred menu names can be predicted. Very serious studies have resulted in menus that increase sales by 30 per cent although the dishes are identical. The icing on the cake is that customers preferred the 'neurologically well-described' dishes.

Another example: in certain high-risk professions such as pilots, saliva measurements aimed at detecting stress hormones can be conducted as part of hiring tests. Saliva analyses are carried out when people are subjected to their future work environment, for example in a flight simulator.

Studies by laser pointer, telemetry, micro-sensors and purchase simulators

Neuromarketing also relies on cutting-edge laser pointer techniques that help track the customer's eyes and accurately measure how long the customer spends looking at every surface.

These techniques make it possible to organize best-selling shelves more systematically. They have also taught us that customers unfamiliar with a salesperson tend to look at his or her watch and shoes more than anything else, as they are probably looking for tangible and easily recognizable elements to help them out of their uncomfortable situation of uncertainty, rapidly prejudging the person's quality. For the brain, as in popular wisdom, the first impression is what really counts.

Telemetry is also used to detect the faintest variations in the turgescence or humidification of the eyes in response to stimuli. The cultural and artistic industry has learned to test its future products by measuring eye humidification and tear secretion. Crying is often a good sign for the future book, film or painting, but before we really cry, which is often forbidden by our prefrontal area, our eyes slightly moisten. It is not visible but it is clearly measurable. There is no conscious barrier; the decision is made instantly.

Wouldn't asking test customers directly how they feel be easier and cheaper? Not at all: a lengthy survey is more expensive than the equipment, and the results are not as accurate. Customers lie more to those carrying out surveys than to themselves. Above all, tear moistening in response to a moving work of art is often preconscious, therefore unknown or denied by the customers themselves. While genius makes some people cry, Neuromarketing moistens the eyes of most people.

Like eye moistening, the analysis of the face's muscular reactions can yield valuable information. Suppose you want to perform a Neuromarketing test without having the budget necessary for cerebral imaging. No problem. You do not need heavy machinery to know whether you can move your customers during a test. A simple surveillance camera and facial analysis software will do the trick. If your test customer tells you he or she is not surprised, just wait a minute. The customer's mouth is a little too open and his or her eyes a little too wide. You did not see anything, but the facial analysis software detected this to maximum precision. It gives you an 8/10 score for the element of surprise of your new offer. Congratulations. Listen to what your customer is thinking, not what he or she tells your interviewers.

Micro-sensors, dedicated to measuring the heart rate, perspiration or skin colour, are useful tools for assessing a person's stress. Increasing heartbeat, variations in the skin's electric conduction, or skin discoloration can be precisely measured and constitute pertinent stress indicators. The retail industry has learned to measure them in store departments to test the pertinence of the depth of a shelf, the volume and type of music played, the room temperature, etc.

It is in the best interest of the Neuromarketing function to know whether or not the customer is stressed, and if the product, advertising or message increases or reduces the customer's uncertainty level. However, stress is only the end result felt by the customer. Modern techniques go further. They measure the constraint, ie the amount of uncertainty generated by a poster, a package, an instruction manual or a selling technique, by measuring the flow of information, information contradictions, the gaps in the information produced by the design of a store or a shelf, selling arguments, etc. Neuromarketing draws many lessons from this.

Neuromarketing studies strive to determine the 'S point', which represents the ideal level of uncertainty to favour a mental activity. The company can regulate, close to the 'S point', the constraint level of an environment to encourage purchases.

Measuring a target customer's resistance to stress is a standard Neuromarketing technique used to segment a market. The technique is evolving. Purchase environment simulators are now available. Test customers are confronted with virtual stores where all the parameters can be changed rapidly, in particular the constraint, while continuously measuring the stress level and behaviour of the test customer.

Certain commercial websites are equipped with simple expert systems. According to the customer's mouse and keyboard behaviour, and therefore estimated stress, the websites modify the information displayed to keep the customer on the right track at the 'S point' that maximizes their purchase potential.

Etho-marketing studies to detect the customer's primary behaviour

Etho-marketing is part of Neuromarketing. Ethology, the science of animal behaviour, includes a section studying the behaviour of humans as animals. This gives Neuromarketing direct access to the animal subconscious of purchase decisions.

Below our neocortex is a real animal brain, which is always active, particularly when the neocortex is weakened by stress, fear or alcohol or when our animal brain is directly targeted by a selling technique calling for a decision.

This is why studying the behaviour of superior animals is an integral part of Neuromarketing studies. Neuromarketing can predict consumer behaviour by studying the behaviour of superior animals. Customers can often have the same reflexes as these animals in an environment where the human brain is weakened:

1 *Stress:* Too fast or too fragmented, and the human brain hands over the reins to the inferior level, the animal inside us.

2 *Alcohol:* Ethyl alcohol initially attacks the least solid, most recent, most hesitant part of our brain, our human brain.

3 *Fear:* Our animal brain is best equipped to deal with this situation, which was more frequent and more vital when we were animals.

4 *Strong irrationality:* When fed with 'incongruent', unrealistic, mysterious, contradictory information, the human brain hands over the reins.

5 *Very strong authority:* When faced with an influential dictatorship, the neocortex gives up and leaves it to the animal brain to make the decisions.

6 *Frontal lobe atrophy:* Some people's frontal lobes are anatomically small or weakened.

We spend a lot of energy on our growth. We are like prematurely born animals; we are open to learning and inventing, but this makes us extremely fragile for a long time at the beginning of our existence. Our brain is therefore wired for a single objective: grow; develop to reach our safe size at the right moment. Not too early, or we would be inflexible too early. Not too late, or we would remain weak for too long.

We have held on to this animal need in the form of a virtual human need, the need gradually to have more, at the right pace: hence the socio-technical studies on the development of product ranges in harmony with the evolution of the customer's 'size'. Modern customer relationship management (CRM) tells the salesperson when to suggest that the customer moves up the range according to his or her growth profile.

Ergonomic studies

These studies will become a cornerstone of marketing science. Intellectual work ergonomics is the science studying humans' intellectual work to increase productivity, satisfaction and safety. The idea is to adapt work to people, not people to work.

Designing products, messages and selling processes adapted to people, their brains, the way their intelligence works and the way they make decisions is the task of the intellectual work ergonomist. Ergonomics produces standards and recommendations to adapt the products, messages and processes to the customer according to new discoveries in the brain, memory, language and other sciences.

Ergonomists specializing in Neuromarketing audit products and campaigns and make practical recommendations to the marketing department so that products correspond with what we know (or think we know) of how human memory and decision-making processes work. The ergonomic audit of a product strives to adapt the offer to the principal elements likely to influence the brain:

- *Memory:* Is the product adapted to human memory?
- *Decision-making process:* Is the offer adapted to the brain's decision-making process?
- *Stress:* Does the offer keep the customer at the ideal level of uncertainty for the purchase?
- *Emotions:* Is the customer's fear taken into account?
- *Language:* Does the product talk to the brain?
- *Senses (sight, hearing, etc):* Is the penetration multi-sensory?
- *Intelligence:* Does the product relate to the right level of intelligence?

Neuromarketing studies are beginning to feature strongly in major international marketing science journals such as the *Journal of Marketing Research*, a publication of the American Marketing Association. In 2009, for instance, two articles in this prestigious journal dealt with research conducted with fMRI methods (Dietvorst *et al*, 2009; Hedgcock and Rao, 2009). An interesting overview of the current and previous academic research dedicated to the application of neuroscience to marketing can be found in Plassmann, Ramsøy and Milosavljevic (2012).

Basic knowledge to access the secrets of intelligence

Perception of the brain and its functions throughout the history of Western humanity

Over the centuries, the importance of the brain as the centre of reflection, emotions and thoughts has long been underestimated. Many of our ancestors believed that the heart was the source of thought and feelings. Hence the expression still in use to this day is that someone 'has a good heart', when it would be more appropriate to say that the person 'has a good brain'.

Even though brain trepanation dates back to 7000 BC in France, 3000 BC in Africa and 2000 BC in Peru, the brain's importance as an organ for reflection and decision making had not been discovered. One of the oldest civilizations, the Ancient Egyptian, believed that the centre of thoughts and life was the heart. Alcmaeon of Croton, in Greece, *c* 450 BC, was the first Westerner who hypothesized that the brain could be responsible for thoughts and feelings. However, Aristotle continued to claim that the heart was responsible for thoughts. The Romans, for their part, did not leave a significant legacy to medicine and even less knowledge on how the brain operates.

The research of Galen (born AD 129) marked the beginning of a dark era for Western medicine. Under the growing influence of churches, in particular the Catholic Church, body dissection and the study of anatomy were forbidden. Thoughts could only come from the soul, the direct emanation of God, the location of which in the human body remained vague. Thoughts were largely dictated by the dogmas that governed individual behaviour. At this time, a small number of scientists took a very discreet interest in brain functions. In the 16th century, Leonardo da Vinci claimed that the soul was located in the brain and that this organ was responsible for controlling the body. In 17th-century Great Britain, anatomist Thomas Willis, who studied how the brain works, is considered the father of neurology, even though some reserve this title for Jean-Martin Charcot, one century later.

By authorizing the dissection of bodies, the French Revolution helped medicine make significant progress. Studies of the human brain took advantage of this new possibility. In the 19th century, Pierre-Paul Broca, the founder of the French anthropology society, weighed 432 brains of men and women. However, the conclusions he drew from his experiment, published in the 1861 bulletin of the anthropology society, that 'the small size of women's brain attests to their intellectual inferiority' turned out to be misguided. After his death, the brain of the person considered one of the most intelligent in the world, Albert Einstein, was given to science and weighed; it turned out to be closer in weight to a female than a male brain.

In 1848, the famous study of the case of Phineas Gage, a man who had lost all emotion as a result of accident, described by Damasio in his book *Descartes' Error* (1994), helped establish a link between the brain and emotions.

As for medicine in general, the deadly wars that brought bloodshed to the world in the first half of the 20th century constitute a considerable source of progress for medical studies, in particular research on the human brain's functions and the role it plays in people's physical and psychological behaviour. This clearer understanding led to a profound change in mentality. As highlighted by Ginger (2007), 'in the years around 1914, it wasn't uncommon for women to consult their physician if they felt pleasure during sexual intercourse'. At the time, female orgasm was often considered a sign of hysteria and sometimes perversion.

On a different note, the US military were among the world's first armed forces, during the Second World War, to consider that a soldier who 'cracked' under the excessive pressure due to combat could feel severely depressed and deserved to be treated rather than shot for cowardice.

Thanks to the change in mentality associated with major evolutions in medicine, the knowledge acquired on the brain over the past 100 years far outweighs that accumulated for 6,000 years in the Western world. The second half of the 20th century saw spectacular progress in knowledge of how the brain works. This progress is largely due to the increase in the number of inventions in many domains, eg pharmacy, chemistry, medicine, physics, radiology, biology, computer science and psychology, and above all the close collaboration between these different sciences in terms of research.

In pharmacology, Professor Henri Laborit created a psychotropic substance in 1952, Largactil, which paved the way for psyche. It spectacularly improved the treatment of patients suffering from psychiatric diseases, who were until then considered insane and treated by excessively traumatic methods. Other researchers, such as Professor Gowlinsky, improved the efficacy of these medicinal products, which resulted in the appearance of Temesta and Prozac.

Insight into the human mind has made significant progress with the development of the different schools of psychology, psychiatry, psychoanalysis and, more recently, neuropsychology. Beyond Freud and his disciples, numerous authors such as Ornstein (1992), Damasio (1994), Bard and Bard (2002), Montague (2007), Singer (2007), Gazzaniga (2008), Changeux and Garey (2012) and Servan-Schreiber (2012) have significantly contributed to neuroscience.

The use of the MRI or simply the EEG, invented by several US researchers in the late 1970s for the former and 1950 for the latter, made significant contributions on how the brain works. These new processes made it possible to study the brain without having to open the cranium, and to understand its interactions with the human body and how it responds to external stimuli. Other recent progress emanating from the study of deoxyribonucleic acid (DNA), genetic heritage, heredity or the genome added considerably to current knowledge of the brain.

Following applications in psychology, medicine, sexology, etc, re-flecting on the use of neuroscience in marketing seems to be a logical step. Neuromarketing appeared in the early 2000s, as a result of Read Montague's research at Baylor University in Texas.

In 2004, Montague and McLure (see McLure *et al*, 2004), in con-junction with other researchers, conducted a Neuromarketing survey based on the MRI. This survey related to consumers' explicit and implicit preferences for the Pepsi-Cola and Coca-Cola brands. As the results achieved were deemed pertinent compared with traditional market research, researchers and large corporations began to take an interest in this new discipline. In 2002, Patrick Renvoisé and Christophe Morin wrote the first book in this domain. This book was essentially dedicated to how the 'reptilian' brain (the primitive brain within us) works and how its understanding can improve sales. The first large-scale symposiums on this new discipline were organized in the United States in 2005. Extensive research programmes were launched by many researchers, often with the backing of large corporations in their country, including Brian Knutson's research in Stanford, which relates to the prediction of product purchases based on MRI analyses. On the other side of the Atlantic, Fabio Fabiloni in Rome, Arnaud Pètre and Patrick Georges in Belgium, Olivier Oullier in France, among others, embarked on Neuromarketing research in conjunction with companies. In Asia, Zhejiang University, located in

Hanzhou, China, also launched Neuromarketing investigations on brand perception. Among the most significant research undertaken over the past few years, it is impossible not to mention once again Martin Lindstrom's large-scale survey on 2,081 volunteers subjected to the MRI in five countries (the United States, the UK, Germany, China and Japan). The interesting results achieved after several years, and the companies that took part in this research, feature prominently in his book *Buyology* (2010), first published in 2008, which became an international best-seller. Since then, many other books, including Weinschenk (2009), Pradeep (2010), Zurawicki (2010), du Plessis (2011), Dooley (2012) and Steidl (2012), have been published, reflecting a growing interest in the field. Filmmaker Laurence Serfati, in France, made an excellent film showing the world's principal Neuromarketing experiments, evocatively entitled *Des citoyens sous influence* (Citizens under the influence). Long excerpts of this film can be seen on YouTube or Google.

To respond to businesses' growing interest in Neuromarketing, companies specializing in this domain are emerging throughout the world, the oldest of which is probably Neurosense, founded by Gemma Calvert in Oxford, UK, and one of the best known of which is NeuroFocus, founded in Berkeley, United States, and taken over by Nielsen. Its CEO claims he is able to anticipate the future success or failure of an advertising campaign based on the MRI observation of a combination of brain responses of a target audience.

Many companies from various business sectors such as consumer goods, luxury goods, retail, hospitality, services, pharmaceuticals, banking and insurance are currently using Neuromarketing techniques. In the United States and the English-speaking world, these companies are happy to communicate extensively on their experimentations. In Europe, in particular in Latin countries, communication in this domain is far more cautious.

The triune brain and the transmission of information

In an interesting article, Rava-Reny (2003) presents how the triune brain works according to the theory elaborated in 1962 by Dr Paul MacLean (1990), from the University of Bethesda in the United

States. MacLean shows that there is one brain divided into three sections, called the trinitarian or triune brain. Each level of this brain has its own characteristics.

Below is a five-point description of the human brain and how it works according to the theory developed by Paul MacLean:

1 *The reptilian brain.* This is similar to the brain of reptiles and dates back to the era of our reptilian ancestors. The primary task of the reptilian brain is to guarantee the survival of our body: drink, eat, sleep, defend the territory (aggressiveness), and ensure survival of the species (reproduce). It is conservative and has the instinct of imitation. It favours the sense of smell over all other senses. When it acts, its actions are knee-jerk, instinctive and quick. They are fairly predictable. This is the first brain system to be formed in human beings.

2 *The limbic brain is the emotions system.* It splits the world in two: 'I like' and 'I don't like'. What is pleasant is registered as to be repeated. What is unpleasant is registered as to be avoided. Inherited from the first mammals, it enables feelings. It helps in caring for children, and having a sense of family and clan. It compares everything with actual experience. It favours hearing over all other senses.

3 *The neocortex or cortical brain.* It analyses, anticipates and makes decisions. It favours sight over all other senses. Relatively bereft of emotions, it acts like a computer. It helps us reason and anticipate the future. In a nutshell, the neocortex is what makes us intelligent.

4 *The frontal lobes of the neocortex.* These sections make the neocortex human. Certain scientists consider them a fourth brain. They enable human beings to be altruistic, to think of others, to create and to embrace the future. The neocortex and frontal lobes make human beings different from other animals.

5 *How does the information circulate in the brain?* The information arrives via the reptilian brain. If the human being's survival or basic needs (food, reproduction and defence) are not under threat, it passes the information on to the next level:

the limbic system. However, in some cases (stress, alcohol, fear, strong irritation, strong irrationality, strong authority, frontal lobe atrophy), the reptilian brain has to make decisions itself. The limbic system assesses whether the information is pleasant or unpleasant. If it is pleasant, it passes it on to the neocortex, which will process it intelligently and often positively. If it is unpleasant, it does not pass it on. The person is then confronted with a negative perception and tends to dwell on it and brood. Therefore, as pointed out by Rava-Reny (2003), people should put a positive spin on situations if they want to 'win', as not only does the limbic brain pass on the information to the neocortex, but the neocortex processes it as a priority. As the old limbic saying goes, 'If you start off as a loser, you will surely lose; if you want to win, act as a winner.'

These reflections on the functioning of the brain can provide Neuro-marketing with a wealth of findings in different domains, such as market segmentation, the management of sales forces, the attitude towards customers, and communication.

Prefrontal cortex and postfrontal cortex: emotional intelligence and rational intelligence

The prefrontal emotional cortex is the youngest, most open and most adaptable area of the brain. It generates emotions and stress by producing hormones in the body to cope with incoherence and uncertainty. The prefrontal cortex produces intuitions, feelings, more intelligent subconscious decisions and more accurate premonitions than the postfrontal rational cortex, which can rely only on ration-ality, calculation, reflection, and extensive and conscious analysis.

If the situation is simple, the postfrontal cortex works alone by calculation. If the situation is complex, the prefrontal cortex has to give its opinion, as the postfrontal cortex is overwhelmed. The pre-frontal emotional cortex is more 'intelligent', as it can integrate more elements than the postfrontal cortex by sampling and approxima-tion. It can perceive things that logic cannot. The prefrontal cortex detects inconsistencies, irrationality, and the inappropriate elements of an offer or product. In a higher position, it warns the customer's postfrontal cortex about a danger via stress. People whose rational postfrontal cortex is predominant tend to say 'Music is what I like.' The object is followed by the feeling. People whose emotional prefrontal cortex is predominant tend to say 'I like music.' The feeling is followed by the object.

For customer segmentation, most companies classify their core target according to its emotion/reason position. The way a product is presented, its design and its offer will vary considerably depending on whether the standard customer is more emotional or more rational.

Customer mimesis: neurons mirroring the premotor cortex (Rizzolatti et al, 2008)

The premotor cortex is the area of the brain between decision and action. It prepares the action, the rest of the movements required, in anticipation of the order to move. The brain shows the film of the action to test and prepare it, like the officer's briefing before the attack. It moves the troops on a map as a dry run.

This area has a strong tendency to imitate the gestures of others around it, to follow suit and to imitate. It has a chameleon's instinct, a herd instinct.

Normally, this area lights up when you make a decision to prepare the battlefield in anticipation of its application. However, marketing science sometimes finds a short circuit, a way to light it up when no firm decision has been made. If it lights up without any specific purpose, it has to do with 'chameleon-like', gregarious behaviour. You take on the colour of your environment. You prepare a plan dictated by your environment, not by your decisions. For instance, as part of the fake customer experiment implemented by the store to make other customers believe the fake customer is buying the promotional products next to you, your premotor cortex lights up, not because of your decision to purchase, as you were still hesitating, but because of a signal from your environment. A dominant decision maker next to you firmly grabs the product and confidently walks to the checkout. Unsure of your own decision, you imitate the person. You grab and buy, relieved of all the stress of having to decide on your own. You are uncertain, so you might as well do what the others are doing: you follow the herd.

Your premotor area will activate the decision orders imported by marketing techniques rather than your internal decision orders if:

- your premotor area is preheated: you are about to make a decision;

- your premotor area is awaiting a clear order: you are still hesitating;

- your premotor area sees a ready-made plan within its reach, ready to be activated, for example the behaviour of a fake customer.

For your premotor area, the external plan is as valid as the plan it expects to receive from your own decision-making centre: the fake customer shows signs of dominance and determination that are the opposite of the uncertainty felt for far too long.

Studies can show this premotor area at work. For instance, in monkeys whose premotor cortex has been equipped with electrodes, it is activated if the monkey sees a movement that it can imitate, even though it does not move yet. It is learning the movement while not yet performing it.

On the fMRI, the premotor cortex of someone who performs an action, envisages performing it or sees someone performing it also lights up.

This can also be observed when studying a spectator watching a footballer running. Even though the spectator obviously cannot run with the footballer, this is not the case for his or her premotor cortex, which is indeed running. If it overheats and its electric activity spills over into the adjacent motor area, this spectator ends up kicking the ball, even from his or her armchair. Indeed, when a person or an animal mentally prepares to act, the electric hyperactivity produced by the premotor area always slightly spills over into the adjacent motor area: the foot or the tail starts moving involuntarily. Closely observing the movements of a good player is very useful for learning a sport, even if you do nothing. Your premotor cortex is already recording the images of the right movements. When you start actually to kick the ball, your brain already has a pre-registered model in its premotor area. Seeing how it is done and doing it are two closely related and interchangeable activities in the brain.

Imitation, eg buying what others are buying, is driven by the customer's stress. In times of crisis, the best thing to do is choose the same solution as others rather than being creative. There is no time to be yourself; you have to make do with what the others do. Blending in is surviving. Mass fashion stores, with their overexcited atmosphere, are managed by marketing techniques to create the stress conducive to imitation purchases. When stress is in sight, the herd closes in. Everyone pulls in the same direction.

When we see people suffer, we suffer with them, or almost. When we see someone scared, we become scared; our premotor cortex lights up our fear area, albeit 2 to 10 times less than if we were actually in this situation ourselves. This level of involvement, or of empathy, depends on the strength of our prefrontal area, which decides whether or not to let our premotor cortex live its imaginary life.

If marketing wants the customer's imitation area to take control, it must first disorient the frontal lobe, and reduce the rationality of the situation. A price inconsistent with the object sold is a good example, often used in marketing, to scramble the filters of the prefrontal lobe and make the customer opt for an imitation choice.

The imitation area is activated when novels, films or characters manage to get you involved or transport you, and when you identify with them. The premotor cortex does not have to see an action in order to imitate it. It can also read or hear its description. At the end of a good western movie or book, after riding with the hero, all readers or spectators walk a few steps with their legs slightly bowed out of empathy with the cowboy.

We buy what others buy, particularly if these others are our role models and our dominants. While companies have not waited for neurologists to get their products endorsed by celebrities, Neuromarketing helps refine the process. If celebrities are too different from the customer, too beautiful or too rich, they overshadow the product, and the customer rejects the product as being designed 'for them, therefore not me'. In this case, should the prescribers be ordinary people? 'If someone who looks like me buys it, it means that I can buy it too.' The downside is that 'This product will not elevate me.'

Commercial efficacy lies with prescribers who are like 'a slightly better version of us'. The fine brain images of Neuromarketing help find product prescribers who light up both recognition and similarity areas ('it's like me') and promising areas ('just a little better').

The drugs in the brain that trigger action

Three hormonal levels

At the base of the brain are chemical 'plants' producing drugs that push us to adopt certain types of behaviour. To put it simply, there are three categories of drugs:

- those that push us to survive, eat, drink, protect ourselves and reproduce;
- those that push us to love, to want to be loved, to be similar, to fit into a class or a community, and to integrate; and conversely
- those that push us to be different, respected individuals.

These drugs are produced in varying strengths depending on the individual. Some people primarily need to be loved and to be similar, while others need to be respected, different and solitary. This is the richness of nature.

Marketing must take these basic needs into account even if they are contradictory. Segmentation into 'social classes' helps predict the strongest needs in a given population. Dopamine is the powerful hormone that signals a possible reward. It increases when we anticipate a higher social status, and when we are about to buy what will give us this status. We respond well to dopamine, particularly if we are socially unsure. Fashion could be the teenagers' dopamine 'dealer'.

Objects representing social status make us produce dopamine. Their possession means more food and sex, as those with the highest social standing are given priority on these means of survival, at least in theory. A specific area of the brain lights up when we see social, fashionable objects the possession of which could lead us to meet more people, to be accepted and loved by them, to join the club and therefore to increase our chances of reproduction.

Brain shortcuts

The world is too complex for our brain, which can survive only by simplifying things. To do this, as demonstrated by numerous scientific publications, it uses prejudices, categories, generalizations and shortcuts, which reduces its intelligence but increases its efficacy, in light of the increased speed of the decision-making process. These mental reflexes are acquired through genetics and learning. Different is dangerous; beautiful is good; if it is red and in the oven, it will burn... These instinctive simplifying associations are powerful decision-making aid tools that can be used in marketing.

Marketing links the brand to a positive symbol for the product: fashion shops in Japan have Parisian names; a technique will be better perceived if it has a German-sounding name; the same is true of cheese with a French name or clothes and shoes with an Italian or English name.

What symbolizes your product's strengths, what has positive connotations and what is easy to understand, even if this has nothing to do with the product? An animal, a country of origin, a specific shape?

To ensure that the brain adopts the shortcut corresponding with the product, the shortcut must be driven home. Repetition in terms of communication is a prerequisite for its efficiency.

Provoking, surprising, exaggerating, shocking a little, scaring: emotions open the doors of memory. Fear is often used to anchor the shortcut: fear of being alone, or of being sick if we do not own the product. Fear is frequently used in campaigns targeting, for example, road safety, the fight against tobacco or alcoholism, or the prevention of certain diseases.

The brain's primary behaviour and its influence on decision making

The pitfalls of intelligence

Georges (2004b) highlights certain intelligence weaknesses that marketing could exploit:

- Attention and perception are limited. You cannot do two things well at the same time, eg thinking and deciding. The salesperson tries to take advantage of this weakness by pushing the customer to make an instant purchase.
- Short-term memory is limited. Presentations must be designed so that the audience retains key points.
- Language makes communication possible, but it also distorts it. We must not hesitate to ask our interlocutors to repeat what they have understood.
- Our brain can only process one-fifth of the information it receives. Judgements and decisions can be distorted.
- We all have two forms of intelligence.

Firstly, we have an ancestral, reflexive, rapid and automatic form of intelligence, governed by nine simple rules:

1 Beautiful is good. We intuitively judge beautiful, voluble, tall and thin people as more intelligent.

2 Different is dangerous. We distrust what we do not know.

3 To reproduce, women will subconsciously prefer men with a flat stomach. Their subconscious dictates that these men are strong and will protect them better. Men tend to favour rotund women. They will see wide hips and ample bosoms as guaranteeing easy and well-fed offspring.

4 The more visible something is, the more important it is deemed.

5 The more permanent something is, the more important it is deemed.

6 The bigger something is, the more important it is deemed.

7 The more repeated and frequent something is, the truer it is deemed.

8 The more accessible something is, the less important it is deemed.

9 What comes first is considered important.

Secondly, we also have a finer, slower form of intelligence, which, after analysis, can tell us the opposite of what our automatic intelligence tells us, generating potential internal conflicts.

Our environment can favour the use of either of these forms.

We activate our fine intelligence when we have the time and when the environment is conducive to this type of intelligence. All it takes sometimes is to increase the speed of information and the level of stress for a person who was in 'fine judgement' mode to switch to 'automatic intelligence' mode. Understanding this is essential for understanding one's intelligence and that of others.

Certain factors can favour automatic intelligence reflexes: stress, fear, strong irrationality, very strong authority and frontal lobe atrophy. In addition, our brain is sensitive to the stimuli produced by our senses: smells, music, shapes, weight, etc. They can contribute to increasing sales in distribution channels (see Part III).

The six stimuli of the marketing decision

Other authors (Renvoisé and Morin, 2002) whose reflection was based on the work of experts in neuroscience such as Robert Ornstein, Leslie Hart, Bert Decker or Joseph Ledoux highlight six stimuli that emanate from what they call the primitive or reptilian brain and influence the marketing decision:

- *Egocentricity:* The primitive brain is egocentric. It is interested in and feels sympathy for only what directly concerns its well-being and survival.

- *Contrast:* The primitive brain is sensitive to contrasts. Opposites enable it to make rapid decisions without risk.

- *Tangibility:* The primitive brain likes tangible information. It constantly searches for familiar and friendly, rapidly recognizable, concrete and unchanging elements. It appreciates concrete and simple ideas that are easy to grasp.

- *The beginning and the end:* The primitive brain remembers the beginning and the end of an event but forgets almost everything in between. This limited attention span has a significant impact on how to present a project.

- *Visualization:* The primitive brain is visually oriented. The optic nerve is physically connected to the primitive brain and passes on 25 times more information than the auditory nerve. The visual channel provides a rapid and efficient connection to the real decision maker.

- *Emotional nature:* The primitive brain reacts strongly to emotions.

Homeostasis

To close this chapter, let us present briefly homeostasis, from the Greek meaning 'remaining constant'. Homeostasis is the ability to retain operating balance in spite of external constraints. According to Claude Bernard, 'homeostasis is the dynamic balance which keeps us alive' (in Bateson, 1979). Any living system must simultaneously satisfy its need for stability and movement to stay alive. If marketing managers want to transform and influence the environment, they must transform this balance. Those who manage to do so are those who can transform their homeostatic balance in anticipation of their market's needs. However, to avoid too much stress for those concerned, they must prepare for and drive change in the event of innovations. The brain attaches a lot of importance to what it perceives as harmony versus imbalance to reduce stress.

03 Neuromarketing in question

Neuromarketing, which is a virtual customer approach method based on the study of the brain, raises a lot of issues for users, legislators, consumer protection organizations and, above all, customers and marketers. In this chapter, we will attempt to answer the questions most frequently asked by customers, as well as companies, and in particular those relating to its legitimacy, relevance or ethics. For example, is Neuromarketing progressively replacing traditional market surveys? Does it have limitations? And on which marketing level can it have a decisive influence?

Neuromarketing and issues raised

What is Neuromarketing?

Neuromarketing, contrary to what some would have us believe, is not a science. It is only an intelligent, focused, marketing-oriented interpretation of major scientific texts on how the brain works. It is the knowledge of the brain's information-processing mechanisms that could inspire the companies whose business is to communicate with the brain, that is to say all companies.

If you want to get into Neuromarketing, read Neuromarketing pioneers: neurologists Read Montague, Steven Pinker and Antonio Damasio, or current gurus such as Martin Lindstrom, considered by *Forbes* magazine as one of the planet's hundred most influential people. They all write books (most of which are listed in the References section of this book) with titles such as 'How do we decide?' or 'Why do we buy what we buy?' Read them, keeping in mind your questions concerning a company that must satisfy the customer's brain, and you will enter the realm of Neuromarketing. As these books are somewhat technical and cover a multitude of domains still irrelevant to the company, we have read them for you, extracting from this ocean of neurological knowledge the pearls that will make marketers happy.

We do not present these recommendations in the form of 'scientific commandments', as we have not yet reached that point, but far more modestly, in the form of new ideas to help your creativity in product and service innovation, in developing a brand appealing to customers, in improving one-to-one sales in retail outlets, in designing communication campaigns, in optimizing the internet tool, etc.

This knowledge of the human decision-making process simply generates sensible ideas, not guarantees, for salespersons and marketers. The term 'Neuromarketing' will probably disappear in a few years, as it will be amalgamated into basic marketing.

How to find which part of this knowledge of the brain could be useful to the company

It's easy: put a neurosurgeon at the same table as two marketing professionals. This book is a dialogue between a brain specialist and two marketing professors.

The brain specialist will talk about frontal lobes, dopamine and so on, while the marketing professors will bring this back to topics more relevant to the reader: How do we decide? Why do we buy what we buy? Why do we remember certain things and not others? How can we write so that the brain can clearly understand?

The brain specialist will not have all the answers but will certainly provide leads, which will result in ideas for new messages, new selling methods, and new products better adapted to the customer's intelligence, which is never bad for sales. The marketing professors can use tools to assess how to transform the marketing discipline and adapt it to the Neuromarketing approaches.

How can companies benefit from neuroscience?

We have increasing knowledge of how our brain and intelligence work, thanks to new techniques and researchers' renewed interest in these last uncharted territories. This knowledge is beneficial to marketing, as investigations not only concern the search for disease treatments but also look at healthy people who must make decisions. Neuroscience can inspire marketers. How do you convince your managers and customers? How do you design a best-selling product? How do you design a communication campaign that will be well memorized and lead to product purchases and brand loyalty?

Are there techniques behind these recommendations?

A technical revolution

As described in Chapter 2, new research and techniques have emerged in different domains:

- In psychophysiology and experimental psychology, there have emerged new clinical experiment methods that are more scientific and statistically valid, and also more efficient field experiments, based on a higher number of people.

- Radiology and neurology provide images of the brain (in the process of making decisions) thanks to progress in radiology and electroencephalography.

- In neurosurgery, direct brain stimulations result in behavioural modifications and help in understanding these modifications. Revelations regarding normal brain function, derived from the neurosurgical treatment of diseases – eg epilepsy, blindness, Parkinson's disease – are emerging.

- In biochemistry, chemical dosages of hormones that motivate us are developed and measured in our blood, saliva and urine.

- Computer science sees the rise of customer behaviour models established by computers, expert purchasing decision systems, and sales aids via artificial intelligence.

- In microelectronics, telemetric micro-sensors measure our secretions and how our senses are oriented at all times. They reveal more about our actual thoughts than any of our interview responses.

- In ethology, new discoveries are made in the etho-marketing domain, as the company sometimes targets our animal brain.
- In camera technology, video cameras analyse faces and detect emotions.

These are just a few of the simplest examples.

Is it credible?

Reasonably so, for two reasons: this book's recommendations are based on research published by neuroscience professors from the most prestigious universities, whose books are seen as references. These publications contain bibliographical references to our decision-making processes. But, more importantly, these recommendations have been tested in the field. Thus, the world's leading hotel operator revamped the customer reception of its largest hotel chain; a leading specialist retailer reorganized its stores; a major bank redesigned its websites; and a sales company reviewed its customer approach in light of these recommendations. And it works: it sells, not always but most often.

An increasing number of scientific publications put to better use by companies

There are an increasing number of quality publications on the brain, human intelligence or the decision-making process. These publications are better read and interpreted as part of a marketing and commercial perspective. Doctors in cognitive science can spend their four-year doctoral course publishing 200 pages on the purchasing decision process that nobody will read because their thesis is published in a remote university, in an obscure language, which is not to say it does not have potential value. This is why we decided not to add another scientific article but to make the most of existing publications that have been generally unexploited in a marketing approach.

The new focus of neuroscience research is more relevant to companies

A considerable number of studies relate to healthy individuals as opposed to patients, bearing in mind that health is not just the absence of disease but also covers well-being and intelligence. The commercial use of the results of these new techniques is no longer a taboo. Intellectual work ergonomics is an emerging science aimed at adapting intellectual work to what we know about how the brain functions. It makes it possible to improve intellectual workers' productivity, satisfaction with intellectual work, and security, in particular when faced with stress. Customers can be regarded as intellectual workers who must make purchase decisions. The company's products must be adapted to what we know about their brain's decision-making mechanisms.

Is Neuromarketing ethical?

Consumer associations have been concerned, and rightly so, about Neuromarketing, calling it an invasion of the brain's privacy, a drug, brainwashing or propaganda! Should a code of conduct be respected when using neuroscience for commercial purposes? Yes. For example, seven-day purchase cancellation clauses protect consumers against overly psychological selling methods. Neuromarketing techniques can be subconscious and manipulative. They can be misleading. Consequently, it has ethical obligations. To protect customers, certain commercial practices derived from Neuromarketing should be banned by law.

Research and knowledge must not be halted. Knowledge is not a crime, but the use of knowledge for unethical or criminal purposes must be limited or banned. Censorship must intervene at the marketing application stage, not at the neuroscience level. Neuroscience is only a tool that the company can use to motivate or manipulate customers. What is the difference between motivating and manipulating? While the techniques are the same, motivation means that both parties share the benefits, whereas manipulation means that one of the parties gets everything. Neuroscience, when used ethically, can be beneficial to all, consumers and salespeople alike.

If we examine the example of the ban on 'small print' texts in sales agreements, neuroscience has confirmed, through numerous experiments, the presence of an information-processing rule in the brain: 'The smaller it is, the more insignificant it is.' This rule influences certain people's decision-making process. This is a subconscious rule that is therefore potentially manipulative. Does this mean that cognitive research on the connection between the size of the information and its meaning for human beings should be banned? No, but consumers should be protected against the excessive use of this technique. A judge may rule that, although the customer signed the sales agreement, the information was presented to him or her in such a way that it may have gone unnoticed despite its actual presence.

The real issue is that marketing often already constitutes Neuromarketing, that many advertisements are manipulative, that the photograph of the inside of the car is often taken with a wide-angle lens, that the burger on the poster is often slightly bigger than the one you buy, etc. The issue of ethics has applied to marketing for years. Marketing has been using Neuromarketing for years without acknowledging it. So the ethical issue is not new; it is and should be a concern for all marketing professionals in their everyday decisions.

How can Neuromarketing be beneficial to marketing?

The end of traditional marketing studies?

Not at all. It is simply an additional point of view. Marketing does not have to scan its customers. This is what laboratories are there for, testing standard customers and products and then publishing the results. The only difficulty is that marketing must learn to decipher the jargon of the scientific journals dedicated to cognition.

A growing number of marketing departments in companies pay for private cerebral imaging studies on a panel of test customers before launching a new perfume, a new TV series or a new car. The brain provides the marketing function with much more accurate responses than traditional, interview-based surveys.

Certain marketing departments use costly but accurate functional magnetic resonance imaging (fMRI) before launching expensive products, and resort to electroencephalography, more affordable but less accurate, to test less expensive products.

Cognitive ergonomics is a new science striving to combine all these elements: images, hormones, studies, tests, etc. Cognitive ergonomics can be very useful in marketing. The University of Wisconsin has a good team in this domain. The conclusions of a recent study confirm that the more customers touch an object, the more likely they are to buy it. You must get the customer into the car after spraying the right leather fragrance, play the right music, close the door, which will make the right sound, and let the customer simmer while the purchase percentage increases.

The science that adapts work (sales) to the individual (the customer's brain) to increase its productivity (purchase) and (customer) satisfaction is called positive ergonomics.

Neuromarketing is everywhere

In political marketing, which word or which posture must the candidate use to light up the areas of the brain that predict a positive vote? In the United States, it has been widely documented that the advisers to George W Bush and Barack Obama used Neuromarketing studies in their respective campaigns.

In audiovisual production, does the trailer of this new film light up the areas of the brain that are generally activated in people who have just been shown trailers of the greatest films?

In criminology, liars light up different areas of the brain to those who tell the truth.

There is no 100 per cent accurate prediction but, if the images of the brain go against your intuition or other studies, be alert, as you may be mistaken. Many companies are no longer reluctant to confess that they resort to mind-reading devices or knowledge engineers in their marketing departments. Marketing departments are investing in

Neuromarketing studies because they know that the customer is not free; they can predict customers' decisions in 80 per cent of purchases and want to raise this figure to 95 per cent.

Does Neuromarketing have limitations?

It is theoretically limitless, but let customers believe in their freedom! Make sure you put a different logo on these two identical television sets, so that they can choose! Do your customers decide freely? No, their genetic history and education determine their future. Their environment limits their possibilities.

- The genetic burden. We all carry in ourselves, within every one of our cells, all the experience of the world. Our DNA, educated by evolution, tells us what we should decide. Different is dangerous, beautiful is good, what is repeated is true, and rare is expensive: these are deeply rooted purchase decision rules, which make us predictable. The more familiar neuroscience is with these rules, the greater the influence of marketing on customers.

- The shock of education. Our parents, our schools, society and culture have an effect on the brain when it is at its weakest, from birth. Banks know that the value of money and the notion of risk are entrenched within us before the age of 16. Some private banks provide classes to rich children from the age of eight. The education received by each social class is easily identified. As customers are segmented, they become increasingly predictable.

- The prison of the environment. Once in the store, customers' behaviour is even more predictable. Customers are easily identified via their loyalty cards, and the outgoing shopping basket can be predicted with 80 per cent accuracy. Our environment dictates our behaviour. It dictates it even more so if it is organized for this purpose, with a good understanding of neuroscience.

In practice, there are a lot of limitations:

- While 80 per cent of our purchase decisions are predictable, there is still some leeway. Customers are intelligent and calculating and quickly outwit the traps of Neuromarketing – except perhaps when they believe they are smelling the delicious odour of fresh bread when it is in fact a chemical smell (identified by code FR115 among professionals).

- Neuroscience can make mistakes or lack statistical validity. The statistical significance of certain studies is not always established, owing to the limited number of subjects studied, in which case the hypotheses presented during these studies are not confirmed. However, 'hypothesis' does not mean 'without value'. It means it is an option worth exploring, the challenge of an expert who is well versed and well read in the domain, which is better than nothing.

- The laws protect those most in need. A reflection period, a ban on certain advertisements, and obligations to inform fortunately limit certain Neuromarketing practices.

- Businesses are ethical. A lot of businesses refuse to use overly attractive packaging, shelves giving the best exposure to high-margin products, overly eye-catching stores or overly streamlined products. Businesses only exist because of the customer's brain as it is, with its need for sex, food, dominance, memories or emotions. Neuromarketing has the means to satisfy the brain, but it is up to its sense of ethics to decide whether or not it should. While deceiving customers can have positive effects in the short term, this policy is more often than not catastrophic in the medium and long term. We believe it would be a mistake for Neuromarketing to go down this path.

Neuromarketing applications to the marketing approach

The marketing approach involves two key moments. The first relates to marketing planning and organization. This is called strategic and organizational marketing. The second focuses on guiding decision makers, based on the information policy, and winning over customers via different 'marketing mix' actions. This concerns operational marketing.

During the first, decision-oriented phase, Neuromarketing sheds new light on several levels. It helps make the right marketing decisions by enabling marketers and those in charge of this function to visualize all the necessary information. Neuroscience shows that presenting all the important information within the field of vision of interlocutors considerably improves the pertinence of the decisions, thanks to better interpretation by the brain. As we shall explore in the next chapter, it is used in aircraft cockpits to reduce risks. In the management domain, the benefits of neuroscience have helped create the 'management cockpit', adapted for marketing into the 'marketing cockpit'. This concept also proposes intelligence ergonomics designed to optimize the functioning of brains working together.

Neuromarketing also helps convince interlocutors of the relevance of marketing recommendations. On this subject, while the influence of Neuromarketing is ultimately limited, its benefit is significant with regard to the pertinence of written and oral presentations. It helps adapt the form of the presentation to how the intelligence of the participants works so as to put them in the best possible listening position to memorize the key points. It results in more positive decisions for the presenter (Renvoisé and Morin, 2002), as these decisions seem more logical and pertinent to the brains of the audience.

Marketing organization and efficiency are also strongly influenced by Neuromarketing contributions. To be successful, the marketing function needs to convince sales representatives, plant managers and employees, communicators and various departments involved in its approach of the relevance of its recommendations. This conviction improves when users contribute, from an early stage, to the development of the marketing propositions leading to the recommendations. Efficient forms of marketing organization are now built upon concrete, cross-sectional projects that integrate different end users into the design phase. The contribution of Neuromarketing is apparent on several levels: in the choices of marketing project managers and their training to optimize their performance; and the ability to organize meetings that place the brain in the best possible conditions, making it possible to achieve maximum collective efficiency.

For operational marketing, the contribution of neuroscience tends to focus on consumers and proactive consumers. Brain investigation techniques provide improved knowledge of the customers' behaviour and decision-making processes. The use of techniques enabling the visualization of the brain's reactions or the measurement and analysis of substance secretions sheds new light on this behaviour. There is added significance to this research when analysing the consumer's behaviour in addition to traditional marketing studies. The other contribution of Neuromarketing is to improve innovation and customer relationships based on a more in-depth vision of how customers respond to various commercial stimuli.

The application of Neuromarketing is likely to result in profound modifications in how customer relationships are built in terms of product, distribution, sales, communication policies and so on. This also directly concerns the entire e-marketing domain, which places enormous value on emotional and immediate reactions, based on clicks.

PART I: KEY POINTS

- As a new management discipline, Neuromarketing helps render the marketing approach more efficient. It contributes to improving knowledge of customers by shedding light on their emotions and investigating their intelligence beyond the statements collected by interviewers.

- Its purpose is not to replace traditional marketing studies but to complete them by providing new insight based on the examination of how the brain works. It refines sales and communication pitches by adapting the actions of sales representatives and marketers to the brain's instinctive reflexes. It enhances the marketers' creativity by putting their intelligence in the best possible conditions to innovate. Beyond the customer approach, Neuromarketing helps marketers submit clear and pertinent requests to their executive committee, adapted to how their intelligence and memory work. Doing this facilitates positive decisions in their favour. Finally, it encourages them to drive the changes rendered necessary by the introduction of the 'sense of the customer' in certain companies by limiting employee stress as much as possible.

- Neuromarketing is based on the transposition into marketing of the studies carried out by neuroscience in the medical domain, which are based on knowledge and techniques aimed at understanding the brain's decision-making methods. These techniques include the vision of the areas of the brain that light up in response to marketing, communication or sales stimuli. The resource used is information via nuclear magnetic resonance, achieved by MRI or electrodes.

- The analysis of secreted hormones is an easier method, as it helps in understanding whether the offers, communications, sales pitches, and locations on the shelf bring pleasure to consumers and encourage them to purchase. The use of efficient scanner-based devices should also be mentioned, as they analyse, in a very precise manner, eye moistening, which is sometimes imperceptible and shows that there has been an emotion. Stress and memory tests are also used, as are ethological studies on how superior animals behave, with a view to understanding how the primitive brain works.

- Neuroscience helps us understand how our brain works, in particular its two forms of reflection and decision: the intelligent form and the primary or primitive form.

- The intelligent brain regularly manages human beings. However, it acts like a slow computer. When it is stressed, ie when it receives too much information in one go, its control is taken over by the primary brain or primitive, reptilian or limbic brain, which reacts instinctively. The decisions are made rapidly but often poorly, as they refer to simplistic norms (beautiful is good, big is strong, rare is important, different is dangerous, etc). Human beings cannot simultaneously carefully reflect and make the right decision. Ergonomics allows marketers to organize their time, thought processes and actions to provide their brains and those of their staff with the best possible conditions in which to make decisions and convince their interlocutors: managers, employees and customers.

- Thorough knowledge of how their brain works enables customers to avoid the intelligence traps set by seasoned salespeople, communicators and marketers.

- Knowledge of Neuromarketing can be particularly helpful in ensuring rapid purchases by inhibiting consumers' intelligent defences, ie by targeting their primitive brain. These practices include subliminal communication. However, the excessive use of these techniques, which promote immediate purchases, does not reflect well on Neuromarketing. Customers are intelligent. Once their intelligent brain takes over, they will quickly realize that they have been taken in and will have no further dealings with the salesperson, retailer or brand concerned. The right approach to Neuromarketing, beyond winning over customers, aims at developing their loyalty. It can only be fully efficient in the long term if it is based on strict ethical rules and a genuine code of conduct among marketers.

PART II
Selling the marketing and organization strategy to the brains of managers and employees

Before being presented to the customer, the marketing approach must be validated by the management, who often meet in an executive committee. Marketers (eg product, market or brand managers), for their part, need to get the prior approval of the marketing department. Without this agreement, they cannot access the resources to conquer the markets where they want to place their offer. To achieve this, they must propose an impeccable marketing plan in terms of content.

Our experience shows that the content of a plan, even if more than acceptable in substance, runs significant risks of failure in the absence of a format likely to interest and convince the interlocutors.

Neuromarketing helps present the overall or partial marketing plan for a region, product or market in a format that improves its pertinence in the eyes of the managers. Its purpose is to give the marketing director and marketers the greatest possible chance of increasing the probability of obtaining a positive response as well as the desired resources. The use of presentation methods targeting the intelligence of the members of the executive committee is essential for gaining their approval. The use of rooms designed in the shape of 'marketing cockpits' is a key tool that helps achieve a positive decision. It improves the clarity of presentations and makes it possible to simulate the consequences of the proposed choices for other management domains.

Marketing actions, which intend to bring the company closer to its customers, are sometimes difficult to implement, all the more so when marketers work in companies that place significant emphasis on manufacturing, product invention, productivity gains or finance. In the public sector, the switch from the notion of user to that of customer and then partner is sometimes extremely difficult to drive home when the institutions are being privatized. These changes cannot occur without causing a significant amount of stress in the employees subjected to them. Stress can even result in substantial collateral damage, which may cause illnesses, depression and sometimes suicide, as has been the case in France for Renault and France Telecom. To successfully complete its transformation, the marketing function must take precautions and drive change in good conditions.

The marketing director and the marketers must be able to listen and convince, as these qualities are essential for managing groups or multidisciplinary teams. They must be able to take a number of measures to facilitate the implementation of the required changes while limiting stress.

In the domains covered by Chapters 4 and 5, Neuromarketing constitutes an important support. Neuromarketing becomes increasingly necessary as the motivation imposed by marketing demands changes, as well as the commitment of colleagues and partners. Chapter 4 is dedicated to the methods used to positively convince the executive committee. Chapter 5 focuses on the methods used to increase the efficiency of marketing managers and marketers, to dynamically and effectively drive change, while limiting stress to a tolerable level for the colleagues and employees subject to this transformation.

Selling the recommendations of the marketing plan to the brain of managers

04

Experience shows that the quality of a marketing plan designed to gain the executive committee's approval concerns its format and content in equal measure. However, as Victor Hugo once wrote, 'Form is simply content brought to the surface.' In our consulting activities, we have all too often seen high-quality marketing plans rejected by an executive committee because of a mediocre presentation format. On the other hand, many written reports are never read or even perused by their recipients because of their excessive volume.

To convince an executive committee, a presentation must focus only on key points. The 'marketing cockpit', derived from the 'management cockpit', helps select these key points. To be convincing, a report or presentation must first of all be read or listened to. This is why presentations must comply with certain rules in terms of duration, volume and ergonomics and must be adapted to the functioning of the targets' brains. The targets must be capable of memorizing the principal ideas and quickly understanding the presenter's wishes. Their intelligence must perceive the logic but also the potential gain for them if they give a favourable opinion and grant the resources requested. To achieve this, the presenters must constantly bear in mind that their presentation or written report *should focus on ideas and recommendations likely to arouse the interest of their interlocutors rather than their own.*

At this stage, the fact that the brain is egocentric should not be overlooked. The personal interest of each decision maker should not be ignored, even if the collective interest will ultimately prevail. In this context, a presentation cannot be an 'all-purpose' report; it must accurately anticipate the expectations of the intelligences it is intended for, whether in oral or written form.

Improve the pertinence of the marketing plan for an executive committee: the 'marketing cockpit'

The concept of 'management cockpit' and 'marketing cockpit'

The 'management cockpit' presented in Figure 4.1 arose from the idea that modern managers are increasingly forced to make quick decisions in a complex environment, in a position similar to that of an aircraft pilot. In the event of a problem, they must respond quickly with full knowledge of the facts and must above all avoid the deterioration of one of the company's vital functions. The idea is to provide the executive committee with a dashboard including basic company indicators, similar to that used by an aircraft captain. The management cockpit consists of creating a decision-making room for the senior staff, designed in accordance with the principles of an actual cockpit, the operation of which is based on 'intelligence ergonomics' rules, ie adapting the presentations to how the human brain works. The management cockpit can relate to all management issues, or it can be reserved for a specific domain such as the marketing plan, in which case it becomes a 'marketing cockpit'.

Numerous companies in Europe, such as Unilever, Disney and Belgacom, have implemented management cockpits to facilitate the decision-making process, in particular at management level. Banque Cantonale Vandoise in Switzerland, one of the pioneers, makes good use of this process, based on a well-structured method.

FIGURE 4.1 The management cockpit

The management cockpit room, ready for the management cockpit briefing

SOURCE: P Georges, NET Research, trademark SAP.

Development of the management and marketing cockpit

One of the world's leading IT companies sent its management team to the neurosurgery department of a university hospital in Brussels to attend a two-day seminar on 'How to increase your managerial intelligence'. At the end of the seminar, a manager of this company, head of the sales department, went to the professor and asked to speak to him in private: 'We are facing stiff competition from our major US rival, and our marketing department must change our somewhat inhuman software image. Can the Neuromarketing method help us? Can it create ergonomic packaging for our software to make it more visible and more useful to managers?' This is how the management cockpit was born; it quickly became the interface between managers and their businesses, the most popular and best-selling method. It is currently successfully sold by the Cockpit Group.

The management cockpit helps managers subject to pressure and stress steer their company more simply. This ergonomic interface adapts the business reporting process to what we know about human intelligence, making it easier to understand for the human brain.

It simplifies and stabilizes the information flow to reduce the manager's stress. It divides the information into blocks (walls, questions, knowledge) naturally and easily assimilated by the human brain. It presents the key data that the manager must possess in order to make the right decisions. This 'war room for civilians' scores 80 per cent in the ergonomic audits of the quality of the information presented to decision makers. The information is presented in the form of four major blocks, which correspond with the four natural stages of human intelligence:

1 Will I achieve my objectives (answer on the black wall of the meeting room)?

2 What are the obstacles to these objectives (answer on the red wall)?

3 What resources do I have to address these obstacles (answer on the blue wall)?

4 Do I adapt sufficiently to achieve the objectives by tackling the obstacles while using a minimum amount of resources (answer on the white wall)?

The information is presented in the form of answers to the questions most frequently asked by managers, such as:

- Are our major projects in good health?
- Are we increasing the quality and productivity of our activities?
- Where do we make or lose money?
- Are we improving the quality of our personnel?

This form of business information presentation is more natural, and therefore more efficient, than the traditional accounting presentation. Everything is prioritized and formatted simply and ergonomically.

A visual exploration is arranged into four layers for each group of indicators: 1) coloured lamps, whose symbolism directly and rapidly targets the brain; 2) visual aids for decision making; 3) traditional charts such as those found in all spreadsheet programs; and 4) the figures themselves.

One of the managers' favourite cockpit formats, in addition to that of an aircraft or website cockpit, is the war room displaying decision-making aid information on the wall. Ergonomics is respected in this format. What is important is presented in a larger and more permanent manner than what is less important. The human brain has a strong tendency to believe and be influenced by large and permanent displays on walls. Credibility is greater, as it is perceived as less ephemeral than what features on a computer screen.

The marketing department of the IT company used the Neuro-marketing method for two primary reasons: 1) designing products within its user interface; and 2) conducting a campaign to sell the product internally via the attractive, emotional, entertaining aspect of this new interface.

Several marketing departments are currently equipped with marketing cockpits to take charge of their own management. These marketing cockpits are designed to respond to questions such as:

- Is the marketing department well managed?
- Does it improve the company's performance?
- Are customers satisfied?
- What are our competitors up to?
- Do its actions help increase sales?

Sell to the brain of the executive committee

Sell the oral presentation to the managers' brain

To be efficient, an oral presentation, based on PowerPoint or other applications, must strive to respect certain ergonomic rules resulting from how human intelligence works. Some of these rules are presented in neurolinguistic programming (NLP) approaches, which emanate directly from how the human brain works, and include the following aspects.

Memory is limited. The presentation must be carefully timed. If its duration is not imposed or chosen by the company, it is preferable to limit it to 15–20 minutes. Additional time, if any, can allow the audience to ask questions.

The memory essentially recalls the introduction and the conclusion. The brain can process only one-fifth of the information it receives.

To gain maximum attention, the introduction must be carefully thought out and meticulous. It must present the issues at stake with the subject matter in a clear and precise manner. It must show that these issues are important, even crucial, for the company without hesitating to dramatize them. As inspired by lawyers, the introduction can announce that concrete solutions will be provided at the end of the presentation; it can also offer original and key ideas associated with these solutions.

The purpose of the introduction is immediately to arouse the audience's interest. To do this, the speaker must avoid 'warming up' with platitudes. Instead, the speaker must immediately get to the heart of the matter to gain maximum attention as soon as the presentation begins.

The conclusion will be no less pertinent. Its purpose is to obtain a favourable decision as quickly as possible. The idea is to place the audience at a level of stress that stimulates their automatic intelligence.

It must awaken danger or fear reflexes should the audience fail to make a decision and allocate resources allowing the marketing function to respond immediately. Moreover, it must clearly specify the expectations of the marketing function that require the agreement of the executive committee: approval for a specific action, budget, resources, etc. The conclusion must avoid any speculation on a domain that would distract the interlocutors' attention from a rapid decision. In particular, it must avoid opening a debate, talking about perspective, showing any doubts, making paradoxical choices, or philosophizing.

Between the introduction and the conclusion, the entire demonstration of the marketing plan must respond to several tendencies of the reptilian brain (Renvoisé and Morin, 2002; Lecerf-Thomas, 2009):

- *egocentricity*, by trying to highlight the relevance of the proposed solutions not only for the company but also for the interlocutors, for example increased stock value for managers who are also major shareholders;

- *tangibility*, by backing the recommendations with studies or concrete examples likely to provide facts supporting the ideas offered;

- *emotional appeal*, via contributions targeting, beyond profitability, for example ethics, sustainable development, and social or societal benefits likely to move the audience;

- *homeostasis*, by showing that all the recommendations of the marketing plan are logical and consistent (including studies, diagnosis, prognosis, strategic decisions, resources of the marketing mix and expected results) and that this harmony can only lead to success, as brain memory is higher when faced with a story rather than a list of facts.

Finally, the presentation must reassure the managers, based on criteria designed to control actions over time, making it possible to implement rapid corrections in the event of discrepancies between forecasts and achievements.

To be fully appreciated, the oral presentation should not forget that the primitive brain is visually oriented, receiving 25 times more information from the optic nerve than from the auditory nerve. An oral presentation must pay particular attention to its visual form. It must highlight, as pertinently as possible, the key ideas it wants to convey to the interlocutors. It must not hesitate to use the most effective visualization method, for example one sentence per slide, or simple and pertinent diagrams.

It is imperative that the visual form of the presentation is in harmony with its oral form. Marketing presenters should be careful not to use their own jargon in front of the executive committee. Careful consideration of how words are used and what they mean to every person in the audience can avoid a great deal of confusion, which can lead to misunderstanding and sometimes unnecessary controversy. A pertinent presentation must have a pleasant form while remaining serious. The use of overly sophisticated slides or gadgets is relevant only if they promote the interlocutors' intelligence and impregnate it with a fundamental idea.

Sell the written report to the director's intelligence

In addition to the oral presentation, marketers are often asked to submit a written report to their manager, whose brain, as we have seen, behaves like a slow computer. As every day managers receive far more information from their staff than they can process, their natural reflex is to put aside an overly voluminous report or 'put it at the bottom of the pile'. To overcome this and make sure the reports are

FIGURE 4.2   A one-page business plan

PROBLEM SETTING

- Key figures
- Issues
- Dramatization
- Deadline for the business plan

DIAGNOSTIC

- Internal analysis: main strengths and weaknesses
- External analysis: threats and opportunities

STRATEGIC RECOMMENDATIONS

- Positioning
- Strategy   Be as precise as possible
- Objectives
- Targets

TACTICAL RECOMMENDATIONS: 'marketing mix'

- Additional studies to implement
- Product
- Price
- Place   Be as concrete as possible
- Sales
- Communication
- After-sales

NECESSARY ORGANIZATIONAL CHANGES
Structures – mentalities – information systems

BUSINESS PLAN POTENTIAL
Cost – budget – expected sales and profit increase – return on investment
(amounts and deadlines)

---

read, an increasing number of companies force their employees to comply with standards in terms of report sizes. A short written report is more likely to be read, all the more so if it is only one page long (see Figure 4.2).

In addition to size, a text will be more pertinent, and more 'marketable', if it respects a number of ergonomic rules adapted to the brain's visual system. Neuroscience provides elements that make it possible to answer the question: how does one write a text that sells?

How to write a text that sells

The brain is equipped with a visual system for reading. This system has specific reading characteristics with which the text must comply; otherwise it will not attract attention, will not be read or will be read with no real motivation. Researchers in visual ergonomics study how to prepare a written communication to facilitate its reading by the brain. The purpose of ergonomics is to adapt work to the brain, not the other way round.

Here is a selection of research results that can be of real use:

- *Transform the titles into questions.* Turning the title of the text into a question will make it easier for the reader to recall the text following the question. A question at the beginning of a text triggers more efficient mental work. A title in the form of a question creates uncertainty, which motivates the reader to remove this uncertainty by paying more attention to the text that follows, as it may provide an answer.

- *Write in columns.* Two or three columns per page make it easier to read long texts. Columns of six to eight words are better assimilated by our eyes. If the text covers the length of the page, our eyes must jump several times, which makes things harder for them. From experience, newspapers have used this column technique for years.

- *Write short sentences.* For our brain, simple and easy is good. All word-processing applications include a readability test function, which automatically detects overly long sentences. These sentences must be broken up.

- *Select the typeface.* Certain typefaces have a 'tail' or serif (eg Times Roman), while others do not (eg Helvetica). This is used as a benchmark when the eye scans the line. A typeface with a serif helps speed up the reading. A word in a typeface with no serif breaks the line and makes this word more memorable.

- *A message should be three paragraphs of three sentences each.* Give one piece of information per sentence, one meaning per paragraph and one message per page or screen. The information must be broken down logically, simply and regularly. The information should be pre-divided to make it easier for the brain of the intended reader.

- *White, black and red: for the brain, simplicity pays.* 'Written in black and white' remains the rule. Use only one other colour: red for attention.

- *Avoid asymmetry.* Do not have one section with a title when another has none, one section much longer than another, or one title with a subtitle when another has none. For our ancestral brain, asymmetrical is not good and must therefore be totally rejected.

- *Use contrast.* Use contrast for what you deem genuinely important: bigger, bolder or brighter – but not for more than 10 per cent of the text.

- *Carefully arrange key information.* Elements on the left, at the top and in the middle of the page or screen are more visible.

- *Have clear intentions.* Titles and headings always summarize the subject matter for the reader. Boxes always give them details. The reader's brain must quickly bridge the gap between the form of the text and its content.

- *Start with the conclusion.* You have one minute or one page to convince. The brain gives a certain amount of importance to information according to its sequence. What comes first is most important, which is logical.

- *Write your documents as for websites.* The first part is always the 'home page', with a summary and six links to six sections.

- *Repeat.* Do not hesitate to repeat the information you want the reader to memorize. People tend to consider what is repeated as true.

- *Write only facts.* Avoid ranting or personal remarks unrelated to the content of the text.

Companies have been creating documents and websites that sell for years without the help of neurologists. They have been using visual ergonomics unknowingly for a long time, and ergonomics is often synonymous with common sense.

Increasing the efficiency of marketers' intelligence

05

Marketing development requires renovating the marketing function by changing the role it traditionally occupies in numerous companies. As a fully fledged member of the executive committee, with close ties to management, the marketing function tends to become more strategic. Its operational role diminishes as this discipline becomes decentralized to the sales department and local entities. Marketing evolution means that new links must be created between the manufacturing of products and services in factories, the management of distribution channels, IT and marketing.

Traditionally oriented towards the consumer, who is becoming more proactive, the role and method of marketing will be extended to other partners, eg personnel, shareholders, the environment, and distribution intermediaries. Mentalities must change drastically so that this function can successfully bring the brand closer to customers. Marketing involves reflecting on the required qualities of the staff in this function, from the director to the product, market, distribution channel and brand managers. It calls for the development of participative structures for each project. It demands the close collaboration of all personnel so that this discipline, which consists of bringing the company closer to its customers, can become 'everybody's business'.

The contribution of Neuromarketing affects the marketing organization on three levels. The first is to enable the marketing director and the different marketers to sustain or increase their efficiency within the company. As their role is particularly complex, they must constantly use their brain and intelligence in the best possible performance conditions to ensure their efficiency. The second is to organize constructive, efficient and positive meetings with working groups involving interlocutors with different qualifications, backgrounds, mentalities and expectations. The quality of the meetings organized by the marketing function has a significant impact on its legitimacy within the company. Finally, subscribing to the marketing ethos results in the need to initiate a profound change in mentality in many companies, which are often product-oriented.

The success of marketing stems from how change is initiated and above all organized and supported. This is a particularly complex task in companies where internal concerns still outweigh customer needs.

Neuroscience shows that this complexity, which is subject to the homeostasis phenomenon, can lead to a substantial increase in stress for the employees involved in the necessary change. To achieve a certain serenity during the recommended changes, the marketing function must anticipate and drive the changes it advocates.

Neuroscience to increase the efficiency of marketing managers and employees

The marketing manager and assistant function is particularly stressful. Neuroscience shows that stress is what we feel when we are subject to constraints. A constraint is the amount of uncertainty within our environment. The stress inherent in this discipline is due to the fact that marketing assistants are constantly subjected to the vagaries of an ever-changing internal and external environment. What the company primarily expects of them is that they anticipate this environment with a view to preparing pertinent responses to face this environment at the right time (eg designing the car that will sell in three years, the distribution channel that will be fashionable in the future, the way to identify and address the tribes and communities that will emerge from the internet). The danger is that a stressed marketer quickly becomes less intelligent and is likely to attach as much importance to minor facts and details as to important things. This tendency in the case of stress is inevitable. Marketing managers and assistants must, more than other managers, be fully aware of this. Neuroscience, by providing advice on health and personal work organization, helps improve marketers' efficiency. Neuromarketing must integrate neuroscience recommendations to help the staff in this function enhance their performance. The old philosophical adage '*mens sana in corpore sano*' (a sound mind in a sound body) could become a Neuromarketing motto, making it possible to improve marketers' efficiency by helping them manage their stress.

Good physical health contributes to improving the performance of marketers' intelligence

Being in good physical health is essential for any improvement plan. Without good physical health, it is pointless trying to enhance one's performance.

Poor diet, lack of sleep, excessive alcohol consumption, absence of physical exercise, and failure to look after oneself make it more difficult for marketers to achieve high performance. Marketers must:

- *Organize their diet.* 'Breakfast like a king; lunch like a princess; dine like a monk.' A consistent breakfast, including cereals, provides the carbohydrates required to prevent pre-lunch hypoglycaemia. If lunch is delayed and breakfast was only a cup of coffee, it is more than likely that, by lunchtime, fatigue and loss of intellectual dynamism will be felt, owing to hypoglycaemia, which may affect the marketer's creativity. Because of the effort of digestion, a heavy lunch causes drowsiness, as the blood is redistributed to the digestive tract to the detriment of the brain, which may affect the pertinence of the marketing judgement. The Anglo-American method of a very consistent breakfast and light lunch is therefore more conducive to intellectual work than other nutrition methods.

- *Organize quality sleep.* If you place a notepad on your bedside table before going to sleep and write down your desires, intentions and concerns, you will sleep better. Sleep is a reversible suspension of vigilance. While we sleep, we classify the knowledge that we have acquired in our long-term memory. Sleep is therefore an important activity, which must be well organized, like the day. Our recently discovered 'clock gene' regulates our natural sleep schedule and duration. There are morning or evening people; there are short or long sleepers. Trying to change is pointless; it is better to naturally discover our rhythm and adapt to it. Always go to bed and get up at the same time, even at weekends. Learn to know your ideal sleep schedule, as determined by the 'clock gene'. Avoid stimulants (coffee, tea and alcohol) before going to bed, as well as all activities that increase the body temperature (eg physical exercise or a heavy meal). Wait at least two hours before going to sleep after these activities. A warm bath and a cup of herbal tea can facilitate sleep, in particular if this is a ritual indicating to the body that it is time to sleep. The sleep rhythm changes with age. After the age of 55, we return to a fragmented sleep pattern, short nights and naps. This is normal; this change should cause no concern. Going-to-sleep insomnia is due to the nervousness generated by the events of the day, while middle-of-the-night insomnia is often a sign of underlying depression. We know from neurologists and popular wisdom that no good marketer should make any major decision until after a good night's sleep.

- *Exercise the muscles, as this is important even for intellectual marketers.* Fifteen minutes of physical exercise at the end of every working day is necessary. A balance must be struck between neuronal activity and muscular activity. Failure to voluntarily move one's muscles is equivalent to depriving the brain of any movement. The brain will therefore move the muscles without the person's consent, and not necessarily the muscles whose movement is agreeable: those of the stomach, neck or feet, creating cramps or unexpected changes. This unwanted mobility may perturb the reflection and affect the marketer's pertinence.

- *Drink far less alcohol.* Ethyl alcohol is a well-known toxin to the nervous system, especially when taken abundantly. Social alcoholism, fairly frequent in managers, consists of regularly absorbing the toxin for social purposes or for the exhilaration of occasional consumption, to forget one's worries. Unfortunately, if the toxin intake is more than half a litre of wine per day, or nearly every day, neurological effects will appear, including impairment of judgement.

Good mental health helps increase marketers' resistance to stress

Marketers' modern, continuous, fragmented work consists of making decisions based on the processing of uncertain and constantly evolving information. These conditions are conducive to stress. Stress must be controlled so as not to jeopardize marketers' ability to analyse, choose and make decisions:

- *Organize a personal 20-minute break every day.* Two civilizations known for their composure and self-control, the English and the Japanese, include the same *Zeitgeber* (the German name for this practice) in their culture: the tea ceremony. The tea ceremony is a typical *Zeitgeber*. The idea is to take a break at a fixed time, accompanied by an unchanging and perfectly organized activity, eg drinking tea. This must of course be adapted to one's own culture. The tea ceremony can be replaced by listening to music, reading a few pages of a book, taking a short nap, etc. The break should ideally last 20 minutes so that it is long enough to have the time to cut oneself off from the activity. It is important to stick to the fixed time of the break. Only then can this break act as a *Zeitgeber*, a 'time out' for the brain. It is also important to respect the ceremonial and ritual elements of the break, as they characterize the event and give it strength as a brain pacifier.

- *Have solid personal organization.* It is impossible to change the amount of uncertainty faced by a marketer on a daily basis. This fact is inherent in the marketer's job. To limit uncertainty, which causes stress, marketers must strive to increase certainty through the effective organization of their work, day and agenda. The certainty introduced by effective time management helps combat stress.

- *While marketers have objectives, they must also have limitations.* Marketers are naturally ambitious and want to progress. They want better things, everything all the time, which is fine and necessary if the company is to improve, but can also be dangerous. Stress can kill. While this extreme case is not frequent, a lot of people suffer from excessive desires. There are many examples: a house that is too big, debt, a senior position, and not seeing much of the family.

- *A little common sense for happiness: 'We must desire what we already have, or just a little more.'* It is, however, important to improve and develop, ie desire what we haven't got. The secret is to find a balance, the right rate of progress and stress that allows us the pleasure to advance without suffering too much. In neurology, this balance is called by some professors the 'S' point, between good and bad stress, between desiring nothing and desiring everything. The 'S' point is the ideal quality of uncertainty within the marketers' environment that maximizes their performance. Most professionals set themselves boundaries not to be exceeded, or work principles. They sometimes deviate from these principles, but not often. This is their safeguard against stress, against this part of their ambition that makes them unhappy.

- *Concentration is the best way to increase marketers' intelligence.* Concentrating on a task is the safest way to increase marketers' performance. It is also an excellent way to protect them against stress. Remaining focused on the same thing for more than 20 minutes as if it were the only thing in their life is not easy. They must train themselves and above all have their own work space so as not to be disturbed during this concentration period.

- *To counter stress, marketers must organize a second occupation for themselves.* To avoid the stress resulting from their situation, marketers must rely on three pillars: the job, the family and the second occupation. Professional life and satisfaction depend on many factors that the marketer cannot control, eg sales or the boss's mood. When these factors are positive, marketers are fine; when the factors are not positive, marketers are not fine. The second pillar is the family, other people, friends and acquaintances, heaven or hell. To protect themselves against these sometimes extremely unpleasant roller-coaster rides, most people organize a third pillar to support their life: a second occupation, which they can control much better and where they can seek refuge whenever a storm is raging. This can be a passion, art, religion, sport, a collection or even another job, far less sensitive to external factors than their official job. It must be a real profession and not just a pastime. It must be as independent of money and other people as possible.

- *To combat stress, marketers must be able to eliminate the multiple stimuli that require action on their part.* To achieve this, they must be able to eliminate useless messages and delegate to their staff when their own presence is not absolutely necessary. Studies show that 50 per cent of the electronic, paper or oral messages arriving on a marketer's desk by electronic, paper or oral means are spam or junk mail. They are designed to steal brain time in order to satisfy the purposes of others. They must be eliminated. A second option is to classify the messages that can be delegated to someone more competent or better informed, who can respond to them more rapidly and efficiently. To be efficient and focus on what is important, marketers must improve their elimination method by using a value grid to screen the information received. The filter is essential so as not to drown in the flow of information and to protect oneself against stress. Another attribute is the ability to delegate. Observation has shown that executives paid €200 an hour spent more than half their time performing €50-an-hour tasks. This situation results in reduced efficiency and loss of revenue for the company. Even if it sometimes seems more stressful to delegate than do it yourself, this approach is necessary to improve marketers' efficiency at all hierarchical levels in the marketing function. Marketers must keep in mind that 80 per cent of emergencies are not real emergencies for them but actual emergencies for their colleagues.

Organizing their work environment helps marketers improve their performance

The work environment has a strong influence on performance. A well-organized desk, enabling the brain to work in good conditions, improves marketers' concentration, reflection and efficiency. Rules referred to as 'intelligence ergonomics' facilitate the work of marketers:

- *Increase office lighting on dull days.* This is particularly the case if the marketer is sensitive to seasons and sun variations. A twilight sensor in the office, accompanied by appropriate lighting, will improve mood, morale and productivity.

- *Place important items within your field of vision.* For the brain, what is visible and permanent is important.

- *When marketers are working on an important marketing project, they must protect themselves against interruptions.* When disrupted by a voice, a face or any kind of interruption such as a doorbell or the neighbouring telephone, it takes the brain 20 minutes to refocus. Consequently, an interruption means 20 minutes lost for the marketer. It is often advisable to protect oneself against interruptions for two hours every day to optimize concentration on major projects. To do this, marketers can use an empty room designed for privacy or work from home if they are on their own, switching off the phones.

- *Work on a tidy desk.* The brain subconsciously inspects all the dossiers arranged on the marketer's desk. This leads to a significant loss of concentration on the document it is processing.

- *Dedicate early-morning hours to reflection.* This is when the brain gets the most blood supply and is most favourable to reflection. Marketers wishing to increase their efficiency must avoid reading their mail and e-mails or organizing a meeting when arriving at work. Morning hours are more efficient if they are dedicated to undisturbed reflection. It is preferable to start the day with what is important.

- *Sleep before making a decision.* 'Sleeping on it', as our grandmothers used to say, is also a scientific foundation of Neuromarketing:

  - A decision should not be forcibly made at night. Waiting for the morning to decide means that the quality of the decision will probably be better.

  - The brain works during the night, classifying the information received the day before, rendering it usable by the intelligence. The next morning, the marketer will have more and better classified information to make a better-quality decision.

Using neuroscience to improve the efficiency of collective project meetings

Experts in neuroscience provide a number of practical recommendations for improving meeting efficiency. When in charge of a project group involving staff from other departments, the marketing director can make the most of these recommendations to improve the efficiency of the team work he or she is responsible for coordinating.

Organizing a team room

It is preferable to dedicate a meeting room to the decision-making process for the project team. The marketing director can use the marketing cockpit method adapted to the project. The room will be equipped with all the dashboards permanently displaying all the key information necessary for the decision-making process. The room will create psychological bonds within the team. It helps in responding to certain brain reactions, eg what is visible and permanent is important, and what is measured is done.

- *Provide each member of the project group with individual information management software* to continuously share the principal information. This promotes communication within the group. Communication will be further improved by getting the participants to vote on the important decisions concerning the group.

- *Appoint a co-pilot*, a high-level assistant who can be perceived as legitimate by the group and can replace you when the meeting does not address crucial issues. You must choose someone capable of solving 80 per cent of the problems as well as you can. This will give you extra reflection time.

- *Define precise meeting rules from the beginning* (arrival time, duration of each meeting, interventions, relationships between group members, etc). It is important that the group leader strictly enforces these rules from the very start.

Avoid showing signs of stress when there are difficulties in moving the project forward

In the event of a crisis or uncertain situation, the participants in the project team must look for points of reference: people who can provide certainty. The group leader must not add to the uncertainty by showing signs of stress in his or her behaviour.

You cannot run around in an aircraft; you cannot run around on a boat; you cannot run around in a hospital. If a person in authority starts running, showing signs of nervousness, all the 'passengers' become even more uncertain and nervous, and the situation can degenerate.

Not showing any signs of stress does not mean ignoring the situation; it means remaining calm because you have a plan. With this attitude, marketing managers reinforce their image as a leader within the project team.

Organize leadership

Leadership is the influence we have on others. It is often a natural gift. Training someone in becoming a leader is often a waste of time. The social intelligence of a leader is determined relatively early in an individual's personal history and can seldom improve. Despite all this, the organization of work can offer some solutions to increase influence on a working group.

People's automatic intelligence follows simple rules to process information. Examples of prejudice should be reiterated: beautiful is good, different is dangerous, what is extremely visible is very important, and what is difficult to access must be important. These are rules applied spontaneously in the absence of sufficient information to make a good judgement. Leadership can benefit from these.

Organize your work so as to maximize your visibility and minimize your accessibility (which is totally different from availability). A lot of people will consider you important. You will increase your influence over them.

Influencing is just another technique

It is not necessarily the prerogative of politicians or journalists. To improve their efficiency in managing a project group, marketing managers must be aware of ancestral intelligence processing rules and put them to good use. If they wish to lead and influence people, leaders must take into account the brain's primary apprehensions, which include:

- What is rare and in short supply is desirable.
- The person who wears signs of authority (eg a white coat or a suit and tie) must be obeyed.
- What is similar is credible; different is dangerous.
- If a lot of people do it, I can safely do it too.
- I must be consistent, and coherent with the group; if I commit to something, I must carry it through.
- If I receive something, I must give something in return.

Do not decide on anything important during a meeting

The intelligence of project group leaders is not at its peak during a meeting. Too many voices, faces and distractions disrupt their brain. They should avoid making important decisions without taking the time to reflect upon the matter and sleeping on it for at least one night. They can convey their decision at the next meeting.

Supporting change to prevent stress

Marketing is 'everybody's business'

Marketing management strives to bring the brain of managers and assistants closer to that of customers and partners in general. It can only be fully effective if this desire becomes 'everybody's business'. Any reorganization limited to restructuring the organization chart will be ineffective if it is not accompanied by a profound change in mentalities. This is a long-term endeavour, which requires the commitment of the management team along with bold initiatives designed to combat the burden of traditionalism.

Any attempt to introduce a mindset that focuses on the customer throughout national or international structures first requires decentralizing part of the authority and resources from the head office to the branches or subsidiaries, restructured accordingly.

The in-depth transformation of mentalities cannot be sustainable without mobilizing every kind of personnel intelligence so that marketing can become 'everybody's business' and not just the business of marketing experts or sales staff. Nothing short of the full mobilization of personnel and human relations departments can help achieve such an ambitious objective. This mobilization must result in the emission of specific and concrete signs designed to reward employees moving in the right direction. All choices in terms of recruitment, training and motivation criteria are reoriented accordingly. An audacious internal communication policy, designed to support change, is implemented. These transformations are essential to success, as the management's convictions will be judged by department heads and employees in light of concrete actions rather than mere words. Failure in this respect may lead to widespread scepticism and the annihilation, through opposition to change and 'double-speak', of all the expected positive effects of the new customer-oriented organization.

Change is always difficult to manage, as it involves a significant amount of stress for the brains of those subjected to it. As previously discussed, it can cause considerable damage and even, on rare occasions, depression or death by suicide for employees unable to cope with and manage their stress. Examples of suicide in major corporations such as Renault and, more recently, France Telecom illustrate this danger.

The profound change in mentality brought about by marketing must not be undertaken lightly. Its potential negative effects on people in departments obliged to change their behaviour too drastically must be anticipated and dealt with. *The switch from the notion of user to that of customer and partner can cause significant disruptions in the brain of personnel ill suited to this transformation.*

The marketing function must carry out a preliminary diagnosis of the problems and risks likely to be induced by this transformation. It is essential to listen attentively to the personnel concerned, as well as social partners, before making this diagnosis. It may be necessary to implement an appropriate system to limit stress for the employees' brains within the new organization. The use of external partners, genuine change management professionals, may prove beneficial when the skills of the human resources department (HRD) in this domain are limited.

Conditions for reducing change-related stress: recruitment, training and motivation

Recruitment guided by the marketing perspective

Recruitment is a prerequisite for affirming corporate culture and guaranteeing that newcomers rapidly comply with the fundamental principles that govern the organization. Recruitment is more complex when the criteria selected must reflect the future employees' strong enthusiasm for international customers. Companies as different as Procter & Gamble, Microsoft, Sony, Nike, L'Oréal, Coca-Cola, Siemens, American Express and Citibank put a lot of emphasis on and take great care with the selection of their future employees. These future employees, regardless of their posting, are selected with a view to the indispensable adaptation to an intercultural environment and must show heightened interest in customers. The behavioural, attitude and knowledge tests taken by the applicants are designed to detect qualities and aptitudes in these domains. If, when asked the trick selection question 'In your opinion, who is the most important person in our company?', applicants fail to quickly answer 'The customer', their chances of being recruited are compromised.

In certain companies, international degrees, the command of several languages and a desire for mobility are considered key selection elements. These criteria are often complemented by specific psychological and behavioural qualities, eg a taste for team work as part of multinational teams, a keen interest in customers, a sense of effort, team spirit, a taste for honest and frank working relationships, and the desire to travel, including expatriation. ING Bank in Belgium uses a battery of tests when recruiting new employees, making it possible to select applicants according to criteria that assess how their intelligence will adapt to the institution's technological and industrial evolution.

When future employees are destined to play an important role within the marketing function, other qualities are frequently required. Recruitment agencies are contracted to assess the applicants' aptitudes in these different domains.

Thorough knowledge of how the brain works is extremely useful for marketing managers, as it helps them develop these qualities within the company. We can only recommend that they read the previous paragraphs carefully.

Neuroscience is not yet widely used in the recruitment of marketing managers and assistants. However, certain techniques are available and can provide more pertinent information than the analysis of the applicant's handwriting or astrology chart, used by certain HRDs for recruitment purposes. In particular, stress resistance studies will help avoid recruiting overly sensitive employees and placing them in a marketing function, which may be dangerous for them. Certain brain lighting tests make it possible to obtain information on the applicant's propensity to innovate and create breakthrough strategies. Companies' current use of these tools is limited, probably because of lack of equipment or for ethical reasons. In light of evolving techniques and progress, the use of neuroscience in the recruitment of marketers is more than likely to feature strongly in tomorrow's Neuromarketing.

Training: a key success factor for driving change

Training is an essential tool for driving change while reducing stress. It helps cope with the brain's natural reactions to the unfamiliar or the unknown, which is how stress is created. Denis Kessler, CEO of SCOR, a world leader in the reinsurance domain, wrote: 'One of the key success factors for the future will be the training efforts implemented to improve productivity and the quality of the services provided.'

Training is an important element that can accelerate change in mentalities. Certain companies consider training an integral part of the organizational process. To be truly effective, it must not be restricted to disseminating technical knowledge but must help in the rapid acquisition of managerial habits to be used in an international environment. It must be designed as part of a medium and long-term plan, the objectives of which are in keeping with the management's wishes and major strategic guidelines determined for the future. Its purpose is to progressively prepare employees' brains for all the changes that will affect the company in the near future. It is also meant to give personnel's intelligence the required flexibility, which will help them adapt quickly to a variety of assignments, in a constantly evolving national or international environment. Designed with this objective in mind, training is considered a trump card for change.

Training must be designed by enlightened managers and provided in a professional manner. Amateurism and superficiality must not be permitted.

Professional training means training that is entrusted to internal or external professionals. It entails the development of a training plan, preceded by an analysis of requirements, as well as exhaustive information on successful initiatives in other companies. It includes a number of well-determined objectives and targets, defines a choice of appropriate resources (educational and financial) and sets up a pertinent result control system. Introducing a marketing mindset in certain companies creates such upheaval among some employees that appropriate training is a necessity.

Training can involve two different albeit complementary aspects. The first relates to driving change itself. As described by Lecerf-Thomas (2009), it enables employees to evolve by 'unlearning' the practices of the past and learning a new method designed to integrate the customer into the heart of the company's concerns. To do this, Lecerf-Thomas suggests following a five-step grieving process: denial, anger, bargaining, depression and acceptance. Training seminars are organized by professionals in this domain. They make extensive use of neuroscience. They can be crucial when change is perceived as particularly stressful by the personnel concerned. The second type of training is more technical. It consists of demystifying the difficulty induced by change and showing the employees' brains that they will be easily able to adapt to a new customer-oriented organization. At this stage, it will be necessary to explain the relevance of marketing for the company and to demystify this discipline, which may seem frightening. Training can be entrusted to external training providers with expertise in marketing; it can also involve the integration of internal marketing staff into seminars.

Adapt personnel's motivations to the organizational issues brought about by marketing

Recruitment and training are crucial foundations when preparing employees for market imperatives. The impact of these initiatives is likely to be limited if they are not followed by the implementation of an adapted motivation system, in particular when the company asserts its desire for globalization. Three domains, which are sometimes overlooked by novice institutions, should not be underestimated. The idea is to prepare the conditions likely to limit stress in the brain of the personnel involved in this significant change:

- Facilitate the move of travelling personnel by offering them a similar lifestyle to what they are used to in their own country. International companies must adopt an international organization to manage their personnel's mobility.

- Offer financial incentives to personnel who agree to be adventurous. International travel too often results in a lower standard of living. This trend is incompatible with a policy aimed at promoting mobility.

- Motivate personnel who travel for their career. In large multinational corporations, promotion within the company is closely linked to the ability to travel abroad.

However, the entire motivation system must be reviewed often by companies if they want to remain durably competitive in tomorrow's environment. They can draw inspiration from the emerging ideas of industrial companies that have heavily invested in this domain. René Robin, Kodak Pathé's Human and Social Development Director, points out that 'it will be necessary to use the full range of skills from all forms of personnel intelligence to improve the companies' economic and social efficiency'. Quoting Michel Crozier, he stresses that 'man is not only the "hand" of the Taylorist system or the "heart" of the 1960s–1970s, but also the "head", ie the intelligence'. To ensure the success of any reorganization inspired by Neuromarketing, the reorganizers must always keep this priority in mind.

Basic psychology has taught us that intelligence is more motivated to fulfil its function when it knows the purpose and reasoning as to why it must be fulfilled. It is even more motivated when it feels genuinely considered and appreciated. Most companies with a modern management structure emphasize this notion. A lot of managers readily claim that the attention paid to personnel, rather than working conditions per se, has the biggest impact on efficiency. The role of managers is to specifically channel and increase the organization's driving forces. During an interview, Akiro Morita, CEO of Sony, criticized US managers for their lack of interest in their personnel. A former CEO of IBM also readily claimed that 'successful companies are those who treat their employees as responsible adults and try to make winners out of them, whereas others tend to treat them as children, based on puerile control criteria'. The assessment systems developed by some companies are riddled with inconsistencies. They demand that risks are taken, but systematically penalize errors and encourage immobility. They want innovation but reward pen pushers. In the name of rationalism, they implement bureaucratic systems seemingly designed to destroy employees' self-image. As demonstrated by the functioning of the triune brain, if you label an employee a loser or a lazy person he or she ends up behaving accordingly. As highlighted by an executive from General Motors:

> Our control systems seem to rely on the assumption that 90% of the employees are lazy, always ready to lie, cheat, steal and embezzle from the company. In fact, we demoralize 95% of the personnel who behave as adults by trying so hard to protect ourselves against the remaining 5% who actually are bad apples.
>
> (Peters and Waterman, 1982)

Any structural reorganization, logical though it may be, inevitably brings about adverse effects. The only way to prevent these effects is to build a mature and positive mindset. Personnel and human relations departments must strive to solve this problem in order to help radically transform the mindset of their institution. The key marketing success factor for the comprehensive adaptation of organizations to their customer's expectations and needs is the strong commitment of all employees' intelligence.

As its primary objective is to adapt the organization to make the most of the employees' intelligence, Neuromarketing largely contributes to any reorganization induced by a focus on the customer.

Internal communication: a key asset to drive change

Internal communication is a weapon of choice for the departments wishing to drive change as a result of a new customer focus. Numerous interesting initiatives are devalued as a result of insufficient communication and the failure to explain their relevance to the intelligence of the employees in charge of their implementation. Internal communication emanates from a close collaboration between the marketing and human relations departments. When designed with this objective in mind, as pointed out by CEO Robert Van Hoofstat, who contributed to its implementation within ING Bank in Belgium, it achieves maximum efficiency.

Training already plays an important role in introducing a new mindset and helping employees acquire effective marketing methods, but this is not enough. Lessons are quickly forgotten if reflexes are not created.

Only the lack of information can allow stubborn bureaucrats to perpetuate their tyrannical managerial methods over their subordinates and employees. Once the information starts flowing broadly and rapidly, their image is compromised so quickly that they must either be replaced or modify their behaviour. This is precisely the role attributed to the information and internal communication department. Its principal function is to ensure the free flow of communication within the institution, from top to bottom, between departments and vice versa. Its objective is also to inform, in good time, the managers of the central and local operational departments of the plan's objectives, make sure they have been understood, respond to criticism of them and explain why certain decisions have been made. Its function is also to explain and discuss potential discrepancies between the objectives of the plans proposed by the departments and those definitively adopted by the management. Finally, it is responsible for centralizing the criticisms and suggestions made at all levels and reporting them to the management and marketing departments concerned.

It maintains regular contact with the internal training department to ensure that the programmes proposed are consistent with all the strategies decided upon. It shares concerns with this department regarding the future, formulated by the management and marketing department, as well as the personnel of head office, networks, subsidiaries, partners and so on.

Some US companies consider that the role played by internal information and communication departments, as well as training departments, is vital. The companies believe that maintaining a spirit of mutual trust at all levels of the organization, and ensuring that most employees feel that they are treated as responsible adults, concerned by the policies adopted, is a necessity. The entire range of internal communication tools, largely based on interactive newsletters and internet messages, must be activated. It must constantly be open to information on employee behaviour provided by neuroscience within the context of Neuromarketing.

PART II: KEY POINTS

- The marketing plan is the backbone of the marketing approach. It proposes recommendations to the executive committee that require its approval before being implemented. It specifies a number of reorganizations as well as the dissemination of a mindset among all employees, making it possible to focus the company on the customer.

- Neuromarketing helps the marketing director in several domains: improving the relevance of the decisions via a pertinent presentation of the main recommendations and resulting indicators; and simulating the potential consequences of marketing and commercial decisions on other management domains (operating statement, balance sheet, IT organization, morale of the personnel involved, etc). The use of a 'marketing cockpit' based on the application of neuroscience, like the cockpit of an aircraft, helps significantly improve the decision-making process.

- Beyond the marketing cockpit, Neuromarketing makes it possible to streamline and give added pertinence to the presentation of the plan by the person in charge. This streamlining consists of designing the presentation for the executive committee by taking into account rules on how the brains of the committee members work. The rules help improve the memorization of the ideas and key messages requiring a decision. They facilitate positive agreement based on a simple, clear, coherent and harmonious presentation, the form of which meets the expectations of the committee members' intelligence. Streamlining advocates oral presentation techniques factoring in temporal and structural constraints so that it can be adapted to the participants' memory capacity. It focuses on the presentation of written reports so that they can be read and understood rapidly and above all interest the recipient because of their pertinence.

- Streamlining concerns the length (a single page) and style of the report, as well as methods used to highlight key points. A pertinent presentation to the executive committee is necessary to obtain an agreement and budget, a prerequisite if marketing is to acquire the resources to win over customers.

- Beyond the written or oral presentation, Neuromarketing helps marketers improve the work of marketing or multidisciplinary teams. By following the rules governing how the brain works, it creates an atmosphere conducive to creative intelligence, which helps make positive and rapid strategic decisions. By adopting cognitive rules in keeping with the way human intelligence thinks, marketers improve their management methods and significantly enhance their efficiency.

- A marketing approach often requires modification of the internal organization to improve its sense of the customer. As with all changes, this modification can cause a lot of stress for employees, in particular those in technical functions, who are not always trained in coping with this type of change. Based on findings drawn from Neuromarketing, marketers can drive this change with increased efficiency, while reducing employee stress.

PART III
Improving the efficiency of the marketing action: the Neuromarketing method

Having sold the recommendations of the marketing plan to the brain of the manager, having driven change and proposed an organization adapted to the 'sense of the customer' while limiting stress, Neuromarketing can now focus on the customer. Based on the 'Neuromarketing method', it will strive to adapt all the elements of the marketing mix to how the intelligence of consumers, or proactive consumers, works.

The Neuromarketing method structures its approach into six stages designed to convince the brain as well as achieve satisfaction with the marketing offers presented. These offers concern all the policies emanating from the marketing mix, ie the product, price, distribution, sales, communication and after-sales. Chapters 6–11 are entirely dedicated to this issue.

Before activating the customer's neurons that control the muscles that open the wallet or purse, you must pass the six litmus tests. This six-stage method will help you overcome the six filters implemented by the brain between the company and the purchase:

1 *Be irresistible.* The first stage is the *poster*, irresistible to the senses it awakens. You must manage to capture the attention and good will of the customer's senses. You must be irresistible to the customer's nose, ears, eyes, etc.

2 *Be remarkable.* The second stage is the *product*, remarkable in the fundamental needs it fulfils. You must give pleasure, dominance and games. This is what customers require to satisfy their two major needs: sex and nourishment.

3 *Be moving.* The third stage is the *offer*, which should be moving for the zones of stress, joy, fear, etc. You must move the customer with a story and aesthetics.

4 *Be unforgettable.* The fourth stage is *sales*, unforgettable by the memories they effect. You must penetrate the customer's memory, with the right language, the right repetitions and the right sequences.

5 *Be beyond suspicion.* The fifth stage is the *brand*, which must be beyond suspicion through leadership, mimesis and shortcuts. You must win over the customer's subconscious decision.

6 *Be irreproachable.* The sixth stage is the *irreproachable company*.

Finally, you must win over customers' conscious, voluntary, intelligent decisions through decision-making aids and accurate segmentation.

At each stage, you will find the answers to the following questions:

- *How?* Simple and practical advice.
- *Why?* Gaining insight into how the customer's brain works will definitely give you ideas.
- *In practice?* Questions to ask, with examples and case studies.

We will conclude this part with Chapters 12 and 13, which are dedicated to concrete Neuromarketing applications.

Be irresistible
Satisfy the customer's senses – Stage 1 of the Neuromarketing method

06

The customer's senses are the doors to his or her brain and purchase decisions. In this respect, special attention must be paid to:

- satisfying the customer's nose;
- satisfying the customer's ears;
- satisfying the customer's eyes;
- satisfying the customer's skin;
- entering through all doors at once.

Neuromarketing must take into account and control the customer's senses. The nose, ears and touch are as important as the eyes, if not more so. These more primary senses give access to less conscious decisions, less controlled by reason. The olfactory nerve, for example, has a direct and priority link to the limbic lobe, our pleasure and memory centre. We must decide in seconds if a smell is good or bad.

Our five senses are like red and humid skin that must be rubbed at the right rhythm to get the customers' attention, followed by their pleasure and memory. The mucous membranes of the nose, eardrum, retina and tongue are holes in our skin giving us direct access to the world that surrounds us. The neurons are arranged in a chequerboard pattern so they can be easily excited by rubbing, contrasts and so on.

Olfactory marketing and auditory marketing are making good progress. Below are a few examples that will certainly give you ideas to satisfy the customer's brain, which likes smells, sounds and so on.

Satisfy the customer's nose

This is easy for bakeries: a little of the chemical smell RV184 blown outside the shop to draw in customers, 'driven by their nose'. Ah, the smell of freshly cooked bread from a chemical firm, delivered in economy five-litre sacks! If you are a publisher, this is also simple. Put a little CJ5 in the pulp, turn a few pages to disperse the product – and the book is sold. The spray behind car dealer counters contains C30, the smell of fresh leather. It even works on cloth seats. Studies show that customers remain inside for longer and buy more. The motto of the chemical firm responsible for all these odours is: 'We make your products irresistible.' And they are right.

Smelling the enemy approaching or rotten food is an absolute priority for our brain. Olfaction is the area of the brain physically closest to the decision-making centres, the most direct route. This is the sense that reason is least likely to contradict. What should a bank smell like? It seems that a smell close to vanilla is most likely to light up the area of our brain that is activated when we feel confident. Vanilla features strongly in breast milk.

As will be detailed in Chapter 13, an increasing number of brands are asking perfumers to create a proprietary 'trademark odour', which will be attached to their stores and which is referred to as an 'olfactory logo'.

Satisfy the customer's ears

Don't forget to turn on the sound

Sounds are important for our emotional decisions. Our brain organizes them thoroughly. Firstly it collects them in the temporal lobes, near our ears, and then it divides the work. The mathematics of music is processed in our left brain, while the harmony of the same piece of music is processed on the right. Each side has its own speciality.

Sounds are the first information to reach our underdeveloped brain in our mother's womb. And the function creates the organ. An academic study has demonstrated that, in children born to mothers who surrounded themselves with music during their pregnancy, the areas of the brain that process music are bigger than those of other children. They will also have an ear for music.

Sounds are also stressors. They must be kept at the right level, the 'S point' that constitutes the ideal amount of stress within our environment to maximize our performance. If the performance requires concentration, the S point should be set at 50 decibels (dB), ie near silence. Conversely, when we need to perform a long and tedious task, we need a little stress to keep us awake. The S point for this task should therefore be higher, at 65 dB, like the ideal music level in stores.

There are different types of sound. Certain sounds are an absolute priority for the brain, like human voices. They are perceived subconsciously from 20 dB. There is a voice radar around us that subconsciously hears and warns our consciousness when something important or dangerous is detected. Our alert brain switches from hearing to listening mode.

Everyone is familiar with the 'cocktail effect'. You are having a chat in a noisy environment without hearing any of the other conversations, or so you think. Suddenly, you become more alert. Somebody has just said your name, three tables away. You immediately start distinctly following this remote conversation that you had not perceived until somebody spoke your name.

Your subconscious listens and warns you, if necessary, of a potential danger.

This is more proof of the importance of our ears. You are at your desk, concentrating intensely. Conversations are going on around you but you do not pay attention to them. If somebody asks you, you say that you don't hear anything, that you are not distracted. Somebody slips in 'Addie won't be here on Wednesday' to the surrounding conversations that you claim you cannot hear. On the

Wednesday, before you could have known it, if somebody asks you whether Addie is here, you will spontaneously answer 'No'. What happened? Your subconscious heard the conversations you claim you did not listen to, analysed them to detect the elements that could concern you and subconsciously stored this information in your memory.

More importantly, if you perceive the sound of human voices around you while you are working on a task that requires concentration, your 'intelligence' can decrease by 30 per cent. This is purely technical. Human voices have absolute processing priority for your brain. As soon as the brain hears a voice, it diverts your blood from the frontal lobe, where it helps you think, to the temporal lobe, which processes voices. You are significantly distracted from any task when people are talking around you. It was common sense; it is now science. Open-plan offices are not conducive to high-level intellectual work. Certain British and US neuroscientists go as far as calling them 'intelligence killers'.

Some recent findings to give you ideas

Neuromarketing works on sounds. What sounds help sell products? What music should be used in stores and advertisements? What sounds light up the pleasure area of the brain on the cerebral imaging?

What marketing knows intuitively has been confirmed. Deep-voiced salespeople or announcers instil more confidence than those with high-pitched voices. Bass tones in animals indicate maturity, dominance, and the one who must be followed to be safe.

Some stores have tested a very special sound, a baby's hunger cry, mixed in with the overall sound system. As a result, women bought more food. The interpretation is that this very special sound awakened a subconscious maternal instinct. Car manufacturers pay ergonomists to study what sound a closing door should make to indicate 'quality'. If you want to sell your spaghetti and save your sauerkraut for later, play Italian music in your store.

Having researched their logo, the brands are now seeking a sound they can call their own. For some of them, like the Windows company, the sound can sometimes help the brand become memorable more than the name or logo does.

Satisfy the customer's eyes

All that glitters is gold. Light is important to our brain; the light we receive is sent to the object and reflected back to our eye, having been modified by the structure of the object.

Our 'pleasure' brain reacts strongly to contrasts and brilliance. A more contrasted, highlighted text is more likely to be read and believed.

Use twilight sensors

People do not buy when they are depressed. They must be a little euphoric to open their wallet. They purchase less on grey, rainy days. If a store is equipped with twilight sensors, on grey days this small device will slightly increase lighting to maintain a stable, subtly uplifting light level.

Why? Chronobiology teaches us that we have two clocks in our brain: one behind our eyes and one in our supraoptic nuclei. One is external, set to what we perceive as bright, ie the sun, thereby following 24-hour cycles. The other is internal and biological. It varies, depending on the individual, between a little less than 24 hours and a little more than 24 hours per cycle.

Newborns who have not yet seen the sun, or have mistaken it for the gynaecologist's lamp that dazzled them at birth, follow a biological rhythm that is not well regulated to day and night. Parents are well aware of this. As the babies grow older, the sun takes over as an indicator of time, prevailing over their internal clock. Children become sociable, which means they finally sleep at night!

What happens, however, if the sun disappears behind the clouds or if it rains all the time? This is a problem for a significant proportion of the population, those sensitive to light and seasons. They switch over to their internal biological clock and progressively shift from the external social clock that everybody around them continues to use. After one week of grey weather, these individuals, sensitive to seasons, are in 'night' brain mode during other people's daytime! As a result, they can feel depressed, and this can even lead to suicide for some of them. This is demonstrated by the fact that the light of the sun (natural or artificial) helps them a lot in phototherapy.

Marketing has always resorted to brilliance to attract customers. In the marketing domain, it is commonplace to ask an ergonomist to accurately measure the brilliance of the shelves, packaging and physical products, and adjust it to the right level to encourage purchases.

The colours of the purchase

Marketing knows its classics: blue and pink for femininity and childhood; red and black for virility; or surprising customers with a traditionally black object in pink.

Here are two tried and tested successes of the ergonomics of colour: 1) in an ergonomic study for a new bank, two colours were selected: one for trust, dark blue, and one for dynamism, yellow; and 2) in an ergonomic study for a website, an all white, black and red colour scheme was chosen.

The three colours white, black and red are extremely sharp. They are always used to increase visibility and attract as much attention as possible, as in the case of road signs. These colours are largely used on national flags. One of the most totalitarian regimes of the 20th century designed its flags and standards based on these three colours. If you want to have maximum visibility at a conference or for a customer, the choice of clothes is simple: black suit, white shirt, red tie or accessories.

Shapes that sell

For the brain, beautiful is good. Give your products the proportions of a perfect face and you will sell more of them: small nose, big wide-set eyes, high forehead, luscious mouth, half-smile. Do you recognize the form? Do you recognize these best-sellers? This is the work of cognitive science. The amygdaloid complex of your brain decides what is beautiful within one second, by subconscious categorization. The first impression lasts.

What is symmetrical is healthy and has good genes. Respect the rules of the three thirds in a package as in the face: forehead; eyes and nose; mouth and chin. You can be neither too simple nor too complex. You must find a balance between figurative and abstract. You must trigger curiosity while remaining understandable and interpretable. Hide higher-order structures in your products, such as trees, repeated patterns, leitmotivs, etc.

Generally speaking, you are trying too hard. Don't forget that the brain can process only one-fifth of the information it receives.

Does your product light up the facial recognition area?

If, during the cerebral imaging process, your product lights up the facial recognition area in your customers' brain, this is a good sign. Any product that the brain associates with a face, because of its curves or shapes, is deemed important. If, on top of this, the shapes of your product correspond with the rules of a friendly face or a baby's face – round, symmetrical, three thirds, smile – you have a winner. Your product is put in the 'important and good' box of the brain's system of prejudices.

The Mini Morris and VW Beetle light up these areas of the brain more than other cars. Is this linked to their success?

If you are designing a costly product such as a new car or a new package, a cerebral imaging study is always a sound investment.

A smiling face in front of us is analysed by our facial recognition area. It divides the face looked at into two areas, top and bottom. If the smile of the person in front of you is disproportionate, ie if the eyes close less than the mouth opens, your facial recognition area issues a wave of antipathy: 'Warning: liar'. A false smile is automatically exposed. Three false smiles from the salesperson and the sale is likely to be lost.

We tend to buy more from smiling salespeople. Our memory rates smiling people better, as it believes they are more likely to help us in the future.

Satisfy the customer's skin

Put some weight behind it

When you see an object, the premotor area of your brain immediately estimates its weight in case you need to pick it up. When you grab the object, if it is significantly heavier or lighter than your brain expected, there is an information divergence alert and the object is put back on to the shelf. This constitutes a considerable handicap for 'self-service' products in the retail industry in light of the fact that nearly 70 per cent of the products taken in hand are purchased.

Here are a few examples. For the premotor area, a quality object must be relatively heavy, which is why Neuromarketing will artificially add weight to the device so that the brain does not light up the 'sensory divergence' area when you grab it. Always make sure test customers handle and lift your future products, even if it is a medicine box. The pharmaceutical industry does this and adds or removes a few grams to satisfy the customer's sense. Certain companies that sell mobile phones or iPad-type tablets do not hesitate to add lead to their products in order to achieve the right weight for the brain.

You would be surprised to realize that the brain always forms an accurate and definitive perception of the 'right' weight of an object, just by seeing the object. It must prepare to take it. If your customers find your future product too heavy or too light, adjust the weight. You will read in the section 'Enter through all doors at once' below why we feel seasick when the information received diverges from that expected. Your products must not make your customers feel nauseous.

Touch your customer

Touching reduces aggression. Touch your customer's shoulder or arm firmly and very briefly. High-quality studies show that this increases sales. Other studies show that customers have a better opinion of the staff if the staff touch them.

Enter through all doors at once

Keep it real

The ultimate weapon is the convergence of the senses. If we see what we smell while hearing it, the brain receives the same information through three different channels. For the brain, this is the ultimate proof of authenticity. Would you like to know a trick to enhance your memory? If you want to memorize a text or figures, say them out loud while you are reading them. The brain receives the same information simultaneously through two independent channels. Therefore it must be true and important, and is memorized.

Why do we feel seasick? Because, when we are at sea, our eyes do not always receive the same information as our ears. Our ears think we are moving, and our eyes, which see only the book they are reading, tell our brain that we are not moving. The brain believes that if such a divergence exists between two senses in the same situation it is because we have gone crazy. And our animal brain believes that we can only go crazy as a result of eating poisoned food. Hence the reflex of ejecting it by vomiting. This is an ancestral reflex that, although it has lost its original purpose, is still active.

Do you want proof of the interdependence of your senses?

Listen to music with your eyes open; then close your eyes. The music is louder, more accurate and more profound. Your brain has become blind and has automatically increased the hearing power to remain safe – without asking you, as usual.

Above all, keep it simple

Your brain can imagine thousands of tastes, sounds, odours and feelings, but always from the mixture of a limited number of basic physical ingredients. For your products, always start with these ingredients, add a few abstractions to stimulate the customer's brain and let the brain imagine the rest. And don't forget the fundamental colours: red, white and black, and then green and blue.

07 **Be remarkable**
Please the customer's brain – Stage 2 of the Neuromarketing method

You have captured the customers' attention by satisfying their senses. To do this, you have chosen from the following ingredients:

- the seven basic odours: camphor, musk, flower, mint, ether, tangy and putrid;
- the four basic flavours: savoury, sweet, bitter and sour;
- the six basic emotions (only one of which is positive!): joy, fear, anger, surprise, disgust and sadness.

Now you must pass your second test: please customers; tell them that better is possible; get them to secrete dopamine and other positive hormones.

In the previous chapter, 'Be irresistible: Satisfy the customer's senses', you learned how to capture the customer's attention and how to be well perceived. You appeared. The customer stopped. He or she did not run away. He or she smelled you ('not an enemy'), heard you ('not an enemy'), saw you ('not an enemy') and touched you ('not an enemy'). The customer stopped, and knows that what you are showing him or her is not bad. It is now up to you to get the customer to come to you, to change direction. Show the customer that what you have for him or her is not only not bad, but even better than what he or she currently has.

How do you let the customer know that your proposition is the best? By showing the customer that, with you, his or her fundamental needs of sex and food will be met – or at least the promise of sex and food through social dominance. Dominant animals have the priority where sex and food are concerned. With your offer, customers will be respected and loved; therefore they will be dominant; therefore they will get sex and food; therefore they will survive; therefore their genes will be passed on, which is what they are programmed or wired for. If Maslow and Darwin say so, we should believe them.

The only purpose of the brain is to please itself

Satisfy fundamental needs

The customer's brain is a learning machine. It learns thanks to reward. If this movement gives me more pleasure, I will continue in this direction. If it gives me less pleasure, I will change direction, like a microbe looking for sugar.

Dopamine, the pleasure drug, encourages us to satisfy our fundamental needs. Happiness is when the duration of all our good moments is greater than that of all our bad moments.

Our three fundamental needs

The base of our brain is occupied by chemical plants, producing hormones that drive us to behave in a certain way. If we move in the direction of these needs, our body doses us with dopamine, the pleasure hormone. If we move in the other direction, there is no dessert for us.

Drug production varies depending on the individual. It is therefore necessary to segment customers by taking into account their fundamental needs. Dopamine increases when we anticipate sex, food or a social status conducive to both. Objects representing social status make us produce dopamine. Their possession means more food and sex, as those with the highest social standing have the priority over these means of survival, at least in theory.

A specific area of the brain lights up when we see social, fashionable objects the possession of which could lead us to meet more people, to be accepted and loved by them, to join the club and therefore to increase our chances of reproduction.

Sex sells

Should sex be used everywhere?

Sex sells. One-fifth of advertisements are closely linked to sex and sensuality. For the human brain, sex and food are the two most important needs, as they guarantee survival: food for short-term individual survival and sex for long-term survival of the species. All products linked to or bearing symbols or promises of the fulfilment of these fundamental needs are a little easier to sell.

The worldwide turnover of the sex industry is higher than that of the motor industry. Marketing is aware of this, and in advertising aimed at male customers it has used beautiful girls to adorn the bonnets of cars for a long time. One of the world's most famous viral videos of these past few years is a film showing Paris Hilton washing a car in extremely suggestive clothes and postures with a view to promoting a largely forgotten fast food brand. One of the major advertising executives of the previous century, Ogilvy, did not hesitate to say: 'If you want to gain maximum exposure for your product, place it next to a beautiful model. If this is not enough, take off her clothes, and if this is still not enough change the model.'

The product must provide the promise of sex to sell more. Look at posters. Sex sells. Owing to self-censorship, marketing often denies itself one of its most powerful selling points.

In the sales domain, the lesson is clear. An attractive and open woman can get contracts signed that nobody else could. In his very interesting novel on negotiation, Walder (1959) shows how the appearance of a beautiful woman, arriving at the right moment, can alter the negotiation process in favour of the party using this woman.

If the customer hesitates between two equivalent offers, a warm and sensual presence can tip the scales in your favour. Recruiting attractive saleswomen increases sales to male customers, and not just in bars. Can you put a little more sex into your campaign or product, while remaining within the norms of course?

Does sex also sell for women? Yes, if the scene shows a couple engaged in a relationship presented as lasting, as revealed by a study of the University of British Columbia. The use of sex in communication is more complex for women, who are guided by their more sensitive left hemisphere, than for men, who are guided by their right hemisphere. While for men the mere vision of a woman's body can create attraction, the vision of a naked man's body often induces a simple smile in women. Conversely, George Clooney in his Nespresso advertisement, although fully clothed, is perceived by some women as particularly erotic.

Touch your customers

Touch is a promise of sex. Recent studies show that salespeople who touch their customers sell more. Touching the forearm is the easiest and most common thing to do.

Does sex seem too vulgar to you?

Show more acceptable 'promises of sex' to your customers, 'predictors' that signify sex in their mind, or scenes of fortuitous intimacy with the opposite sex.

Red lips: a strong signal

We are still animals to a certain extent. Read Desmond Morris's books on animal behaviour. *The Naked Ape* (1999) will teach you as much about marketing as many books on the subject.

This is one of many examples. To attract customers, display red lips on packaging or in the shape of the products. Oddly enough, this will attract not only males. Why? For efficiency purposes, the male must approach the female only when she is fertile. During our animal period, which is still very much alive below our cortex, this was easy to predict. We were walking on all fours, with our genitals in plain sight, and indications of fertility did not go unnoticed. Then our species began to stand up straight. The female, standing on her feet, hides her sexual organs from the male's view. How, in this new position, can she indicate to the male that she is available? By reproducing on the second floor what the first floor promises: by wearing lipstick. In the words of our teenagers: 'Kissing is like asking the second floor whether the first is free.'

Serious studies show that some women wear lipstick only at certain times, without knowing exactly why. In any case, fullness and lip shapes have remained beacons, a Holy Grail lighting up for the customer. Look at the products around you and you will find a lot of these shapes. Is the most famous brand in the world (Coca-Cola) not red, moist, lip-shaped and containing exciting molecules? It is also a well-known fact that faces with thin lips are less attractive than those with sensuous lips.

Sports cars: the promise of sex

Peacocks spread their tails to attract females. Their genetics tells them to propagate their genes and that this is as good a way as any. Sports cars seem to serve the same purpose. Behavioural scientists believe that the woman's brain seeks out a strong and healthy male to help spread her genes and protect her offspring. It is more likely to be the one with a nice sports car. You can only invest in unnecessary objects if you are healthy and strong enough to have met the required needs. This is how the luxury industry was born. If you cannot show your beautiful genes, show your beautiful wheels!

The right shape of the bottle? The shape of fertility

It is the object that must be grabbed from the shelf before all others: a shape reminiscent of the first sculptures of prehistoric men (thin neck, wide chests and hips, thin legs, etc); a brand indelibly associated with survival of the species in our genetic code; a shape that can be prayed to or grabbed. Hide the number 8 in your packaging or advertisement. The customer's eye detects it subconsciously before anything else. It stays on it a few milliseconds too long, time for the product to penetrate the memory. The laser pointers that follow and calculate the time and movement of the customer's eyes on store shelves clearly demonstrate this.

Sex sells, but...

The perverse effect, so to speak, is that sex can distract the attention of the customer, who clearly remembers the image but not the product. It can also repel a few customers with a negative image of sex. Advertisements for bras on bus shelters are very popular among men, who can memorize and describe in detail what they have seen. These advertisements even cause a significant number of urban traffic accidents. However, when the men are asked what brand this advertisement concerns, it is often largely forgotten. Or, worse, it can be attributed to the wrong company, which had nothing to do with the communication. The Obade brand, a leader in the sensual underwear market, is a good example of cannibalization of other lingerie brands.

There are two solutions. The first is softer, more acceptable sex that is better integrated into the product. This is easy for lingerie but more difficult for washing powders, although interesting ideas have been observed in this domain. Or, conversely, be controversial: strong, unforgettable sex, bordering on the intolerable. In other words, go for provocation, bearing in mind that forbidden books are always best-sellers. But, as mentioned above, be careful of bad taste and be aware of the risk of rejection, which can be detrimental to the creation of a desire for the product.

The food that gives pleasure

Should customers be drugged?

Life is too hard; we all take stimulants and mood enhancers. We are hooked on drugs, whether they involve ingesting substances (wine, beer, caffeine, cigarettes, cocaine, heroin, ecstasy, etc) or not (gambling, shopping, stress, compulsive exercise, etc).

With external drugs, we swallow the stimulating molecule. With internal drugs, our behaviour boosts our internal drug-producing plants. In both cases, we overconsume a molecule that renders our brain mad and dependent. Confusing the goal for the reward, the brain can no longer be satisfied and therefore always wants more. The brain is extremely sensitive to the drugs that deceive it by taking the place of its natural hormones. The goal becomes the reward and the brain deviates from its path. It stops learning, thereby limiting itself. It auto-ignites, which is referred to as 'dependency'.

You can drug yourself with or without substances, alcohol or gambling; it is all the same. Whether you ingest physically or virtually, it is the same dependency. The addiction to the internet in some teenagers constitutes a genuine virtual drug.

Companies have been using stimulating drugs to sell for centuries – nicotine in cigarettes, alcohol in drinks, excess sugar in food – to guarantee 'repeat business' and captive customers. Conversely, alcohol-free wine or beer does not sell well.

Peanuts?

Bartenders know that the likelihood of a customer ordering another drink increases significantly if the first drink comes with salted peanuts. Salt makes us thirsty.

The ingredients of certain food products make us hungry by inhibiting the hormones insulin and leptin, which normally tell us when we have had enough to eat.

Researchers from the UT Southwestern Medical Center in the United States have documented the specific role of palmitic acid in the urge to eat. This component can be easily added to many food products, which as a result become irresistible. Biochemists are well aware of this, as many of them work for food companies and participate in marketing meetings. Some of them could not resist the temptation of adding a little palmitic acid to their products. The name on the label is innocuous enough.

Another technique used in a supermarket is for the salesperson to offer you a taste of foie gras before lunch. Your brain, thinking it is time to eat, orders the creation of gastric juices, which increase your hunger. As you do not dare eat the entire tray, you purchase a few boxes, or even more when presented with a gift, as the brain usually wants to give something in return.

Get your customers to play

Lotteries, casinos, slot machines and the stock exchange are extremely profitable businesses. Humans are natural-born gamblers, and poor gamblers to boot. Despite being aware of the fact that their money creates the prosperity of lotteries and casinos, they like it.

The brain, a very poor statistician, always believes it can beat the national lottery or the stock exchange. Gamblers invent a betting system or a technique. They read horse racing magazines and often think they are smarter than a system designed to make a profit and to take their money. Adding a sort of lottery to your products can enhance your customer's attention and the appeal of your product.

Let them dominate

The need to dominate is a fundamental human need, which increases the chances of surviving. Certain offers are designed to let the customer dominate. They try to appear as ideal prey for the predatory customer: slow, weak, attainable. These offers are found in the corner of a brochure or store, somewhat neglected, hidden but not too obviously, undersold: customers believe they have found a gem that others have probably not noticed. They rush in.

In nature, to ensure survival, there are two ways to dominate others: gently, by getting them to like you, or forcibly, by getting them to respect you. Love can be characterized by the secretion of two hormones for the brain: oxytocin for acute love and the luteinizing hormone for chronic love. Oxytocin relates to genital love, the need to reproduce. This hormone has a clear objective: survival of the species. It provides assistance during childbirth. The more of it we have in our brain, the more we make love. It becomes less powerful once it has fulfilled its purpose. When the woman is pregnant, oxytocin diminishes, but nine months later it is back, ready to spread its genes and increase the population of the planet. This does not apply solely to women. You can try to whisper in your customers' ears that your product will make them loved or respected by others.

Faced with your future product, ask yourself:

- If you help your customers obtain it, will they be more respectable?
- Does your product involve authority over others?
- Does it give the sign of authority to those who possess it?

Humour sells

Why do we laugh? To remain dominant in unforeseen situations, so as not to lose face. We laugh at an unexpected punchline. The brain goes 'ha ha ha' to find a way out, to reconcile two conflicting pieces of information. Laughing means 'I understand; I am on top of the situation.' Make your customers laugh or smile, and they will be grateful.

Capturing the customers' senses pleases them. They are full of dopamine. They know that, with you, better is possible. If you wish to go further into the process of selling to the brain, you must move them. How? This is what we will see in the third stage of the Neuromarketing method: be moving. Create emotion by stressing out your customer, up to a point.

Be moving

08

Satisfy customers through their emotions to gain their loyalty and ensure they move up the range – Stage 3 of the Neuromarketing method

Satisfying your customers through their emotions helps you hold on to them before consumption and ensure they return later.

To create emotion in the customer, the following approach can be adopted:

- Manage the customer's emotions.
- Stress the customer, up to a point.
- Show them films taking them on a journey.

In one second, during the first stage, you have captured the customer's attention. You have stopped the customer in his or her tracks. In one minute, during the second stage, you have diverted the customer from his or her path by promising better things. The customer is now in front of you. Get the customer to sit down. You have one hour to move the customer, hold on to the customer, and get the customer to test the emotions of the offer you have managed to sell him or her. You must move the customer during and after the consumption of the product: to gain the customer's loyalty and ensure he or she moves up the range.

Manage the customer's emotions

Brain study and appearance or secretion analysis technologies help in perceiving emotions

Customers decide with their gut. Half of their decisions are irrational and emotional. Marketing must deal with customers' emotions as well as their reasons to buy. It must manage customers' stress, fear, disgust, attraction, joy, sadness, remorse, regret and trust. A thorough knowledge of how the brain works, as described in Part I, is crucial for managing the emotions. We have six basic emotions, each of them forming a characteristic expression on our face. Fear, joy, sadness, surprise, disgust and anger are the muses of marketing, the doors to the customer's memory. All studies conducted on the brain based on the different techniques mentioned above make it possible to significantly improve analysis and understand customers' emotions. The mother of all emotions, ie stress, is measured with increasing reliability.

Awaken stress, the mother of all emotions

Marketing needs to know whether or not customers are stressed, and if the product, advertisement or message increases or reduces their uncertainty level. The stress of an individual is measured with growing accuracy. The heart rate increases, perspiration appears and the skin changes colour. Customer stress is measured remotely, once electrodes have been attached to the customer. The customer will forget about this apparatus and will behave normally. Studying the 'S point', ie the ideal level of uncertainty for stimulating mental activity, is particularly important. This is how the company can accurately regulate the constraint level of an environment to encourage purchases at the S point, the ideal level of uncertainty for this activity. According to Patrick M Georges (2005), 'a store set at 220 constraint points will maximize sales to the category of 20- to 30-year-old women of social level 4, a population whose S point is at this level for this type of activity'.

Neuromarketing, via its commercial approach process, strives to position the offer at the S point. Too little uncertainty makes things boring. Too much uncertainty induces stress, which paralyses the decision. If the standard customer, monitored by electrodes, has shown too much stress within the test purchasing environment, he or she must be reassured and certainty must be reintroduced via the usual techniques: purchase reversibility, purchase compliance, etc. Conversely, if telemetry measurements show that the customer is not excited enough, his or her uncertainty must be increased via other usual techniques: increased choice, quick passage, etc.

It is necessary to avoid stressful incoherence in messages, such as obvious incompatibilities between the object and its price. Failing this, inconsistencies must be explained, eg 'We buy large quantities, which allow us to reduce our costs and provide you with quality at an extremely competitive price.'

It is one thing to manage the constraints of customers as a whole, but you can be more accurate and adapt the constraint to what you know about the stress resistance of your different customer segments. Some of them tend to be 'stress lovers', while others fit into the 'stress avoiders' category.

An offer, a store and its selling process, set at 200 constraint points for a general population with an average stress resistance of 200 points, may be deemed boring by a youngster whose average stress resistance is 220 points, and too stressful by a senior citizen with an average resistance to stress of 180 points.

Various standard customers have different levels of stress resistance, owing to their varying chemical production levels of stress resistance hormones. Customers segmented as 'stress lovers' like stress and buy to feel stressed; it is their drug. 'Shop till you drop.' They account for 10 to 20 per cent of buyers in certain sectors. 'Shopping addicts' are well known by retail companies and are managed in the same way casinos manage gamblers.

Most methods for destabilizing customers by stressing them are well documented. Certain sales representatives do not hesitate to use and sometimes abuse these methods, while customers are unaware of them:

- 'Do you have a booking?' This, pronounced with an annoyed tone, prepares the customer to settle for any restaurant table or hotel room. This is called 'softening the customer', who arrived determined to obtain a good table.

- 'Decide quickly. These are the last ones!' The salesperson accelerates the flow of information to render customers less intelligent and to force them to use quick but low-quality animal reflexes instead of slow but high-quality reflection.

- 'First come, first served.' This is another expression, sometimes used by antique dealers, philatelists or collectable item dealers when dealing with a collector. The collector feels obliged to react very quickly if he or she doesn't want to miss out on the bargain of the century.

The Zara retail chain increases its buyers' stress by refusing to restock on any clothes. If customers do not buy straight away, they are not certain they will find the same item later.

Stress to enhance the marketing performance

The stress hierarchy

Below is a hierarchy of the progressive signs of stress and mental overload in a standard marketing department employee:

- *Step 1:* irritability, overreaction, mood swings and appearance of emotions, even though everyone knows that the reality of everyday life is neither good nor bad and therefore should not generate any emotions.

- *Step 2:* sleep disorders. Sub-steps are classified in order of severity: step 2A, difficulty in falling asleep; step 2B, waking up too early; and finally the most serious step, ie step 2C, waking during the night. Your judgement is already seriously impaired.

- *Step 3:* the medical step, ie anxiety, depression, stomach pains and risk of cardiac injury. You will die on the job. In cynical terms, the problem for your employer is the costs involved in replacing you and the fundamental risk to their image in the case of a suicide. Renault and France Telecom have learned this the hard way. This can even cost the CEO or managing director his or her job if the media and politicians pick up the story.

How does stress express itself in your brain? In a clearly identified part of your brain, a little person is sitting comfortably, carefully watching two screens. The first is that of your eyes, where he or she sees what is actually happening. The second is the screen of your imagination, where he or she sees what you think should happen. If there is a difference between what he or she sees on both screens, the little person pushes a button, which gives you a rush of stress hormones in the blood. He or she helps you cope with this unforeseen event.

Manage stress to improve the performance of sales representatives and communication

Maintain a sales representative at the S point of stress

The challenge is to maximize the performance of sales representatives by improving their stress level. Salespeople generally have large glands just above the kidneys. These produce hormones that help them handle stress well. Salespeople are often 'stress lovers'; otherwise they would not be in this business, where there are many constraints.

How do you keep salespeople dynamic but not too much so? With too little stress, they fall asleep and only go for easy customers. With too much stress, they become too aggressive and no longer participate in the team; they adopt a very short-term approach.

For the remuneration stressor, the S point should be set at two-thirds fixed and one-third variable remuneration, provided the calculation is made on what salespeople fully control, ie their own turnover and not the company's profit, or sometimes the turnover of a team of salespeople under their authority. Fixed remuneration provides standard-of-living security, while variable remuneration provides the energy to do more. But you knew this before you even heard of stress ergonomics.

Split the visual of your communications in two

Once you have completed the design of a poster, advertisement, packaging or message, removing half the words, colours, characters, images, etc will almost always improve the cerebral impact of your message.

The customer's brain is slow and inaccurate, much more so than you think. This slowness is concealed by stress. When you simplify, you reduce the stress, which is a good thing. Campaigns, advertisements and stores are often too stressful.

Ask an intellectual work ergonomist to measure the constraint of your messages, stores, websites or products. Ergonomists are generally well trained and badly paid for this.

Frighten customers and promise them good health

Customers are afraid of death and suffering. Promise them good health.

Your new mobile phone has no anti-Z wave filter? This is a mistake. You should install one. Your customers will be more than willing to invent what it is. Because of a typo, the instructions of a mobile phone featured an anti-Z wave filter for one year. The producer dutifully removed the typo. How wrong this was! Since they heard about the removal of the filter that never existed, customers have complained of headaches when they use the phone. This is a true example to reflect upon.

Surprise them

All offers must include a surprise. A good surprise is better, but simply surprising is good too. The traditional surprise is humour, as mentioned before.

Lightening a normally heavy object is also a surprise, as Apple knows. How about painting a white fridge pink? Seth Godin, a marketing guru, vehemently advocates this idea in his book *Purple Cow: Transform Your Business by Being Remarkable* (2002).

Make a film out of your offers to move the customer

Bring customers on a journey to move them

The foundation of the sale is to make customers tense by creating a necessary level of stress and then bring them on a journey to move them and resolve their stress. To manipulate the customers' emotions, there is no industry like the film industry. Your customers are holding your product. Take them by the hand; tell them a story. Take them to the movies. You have one hour to fill them with joy and emotions.

What moves the brain, as cinema, the emotive industry, fully understands, is stories. An evolving plot, evolving characters and evolving decors create stress from the beginning to the end, where stress is resolved. Storytelling is subject to increasing attention from the marketing and communication domains (Godin, 2005; Clodong and Chétochine, 2010).

Your offer has come to the attention of the customer, who knows it meets his or her needs. He or she will consume. What experience are you now going to offer the customer? Prices, promotions and placement are only the beginning of the story between the customer and your offer. Consumption must end well. The promise of the happy ending, which generates loyalty, must be kept.

Ask yourself the following three questions:

- What characters will symbolize the consumption of my offer?
- What plots will evoke my offer?
- What decors and what circumstances do I want as a backdrop to my offer?

What is the story of your offer? How do you make sure your offer is consumed? Below are real-life examples that you will recognize:

- In a special glass? With friends?
- When the customer is driving your offer, who is he or she?
- Have you told the customer what to do when he or she puts the cup down?

- Where does the farmer he or she helped by buying the product live?

A brand is like a film

The human race is worried. It is impossible to predict the future. Despite all its efforts, the brain is unable to calculate this world or predict it, even though emotions are a helpful addition to its calculators. It does not like anarchy, which is why it will create uncertainty reduction devices from scratch, and invent organizing principles for reassurance purposes (religion, nations, etc), which make the future a little less uncertain. Organizing principles are simple conventions and beliefs that cost nothing to invent but that structure chaos (a god, a king, 10 commandments, a constitution, a church, a hierarchy, rules to be respected, a romanticized story, etc). The excessive flow of information received by our brain is filtered, and we finally know what is God and what is Evil, who to vote for, and so on. The most educated populations try to do without and create their own system of values, which is not always easy. Total freedom means total anxiety. A few drops of belief can make things better.

To gain followers, a brand is built like a nation, with a flag, a code of laws, enemies and buildings. No stone should be left unturned: if a brand is incomplete and lacks one of the following six pillars, representative of a 'mental universe', the bubble bursts. Customers must embrace your universe. The brain must confuse myth and reality.

The six pillars of a 'mental universe' are:

- a simple symbol, as in the 'strength' of an apple or Nike's famous swoosh logo;
- a chosen one, a god made human, eg a cowboy, a Michelin man, a clown or a colonel;
- a code of laws, eg McDonald's code of laws, which is bigger than the bible;
- a legendary story and a secret, eg Coca-Cola's legendary story and secret formula;
- a common enemy, eg obesity, boredom or social isolation;

- rituals, eg a mass, a party, a national anthem or a local culture of belonging.

A well-constructed brand will light up the same areas of the brain as the presentation of a nation or religion during the cerebral imaging process. Martin Lindstrom in his book *Buyology* (2010) illustrates this experience, based on MRI studies carried out on a sample of nuns and brand addicts.

Wish your customer joy and serenity

You have moved your customers. Their memory is at your mercy. The fourth stage of the Neuromarketing method consists of remaining in their memory. Be unforgettable. You have one day to gain loyalty and make customers move up the range. You must confirm their loyalty to enhance your offer.

09

Be unforgettable
Satisfy the customer's memory – Stage 4 of the Neuromarketing method

Pleasure and emotions have opened the doors to your customer's memory. Now you must stay there, because without memory there can be no decision to purchase.

This fourth stage of the Neuromarketing method attempts to:

- increase the customer's memory via repetitions;
- increase the customer's memory via stories;
- increase the customer's memory via pleasure;
- increase the customer's memory via simultaneous and sequenced entries;
- increase your own memory.

Increase your customer's memory

If it is repeated, it is retained and must be true

The method of increasing the memory of customers through repetition is well known to marketers and sales representatives. As set out by Georges (2004a):

Describe the unique benefits of your product in one half-page. Learn them by heart and follow the procedure. I will repeat it to you to be perfectly clear: one hour with the potential customer. Right at the outset, tell him exactly what you have written; repeat it after one minute; repeat it after 10 minutes and repeat it just before leaving him. Send him this text the day after; send him the offer one week later, accompanied by your text, and your first reminder is three weeks later.

What is repeated is true; it's in our genes. What is repeated is retained; it's in our memory. Will these six repetitions of a short text not annoy customers? No. There is a progressively longer interval between them, designed to strike the memory at the right moment: not too early when the memory is still fresh because it is irritating; and not too late when it has already been forgotten, because that is costly.

To leave a lasting impression, there is no secret: you need to repeat. However, repeating is expensive. Therefore you need to repeat a minimum number of times: six times. But repeating a commercial message irritates customers, which is why repetitions need to occur at the right moment. These are ideal frequency conditions recommended by memory ergonomists.

The brain remembers when it is told a story

Increase the customer's memory by using stories:

> A: I pay thousands of euros so that, in your film, Brad shaves his beard with our razor. And our tests tell us that not one moviegoer will remember it!

> B: We will fix this. He will cut himself with your razor, get infected and die.

> A: That's more like it. Do you think it's possible? Will the film not finish too early?

Is product placement a costly lottery? A little less so since we have discovered how human memory works. All companies should subscribe to scientific journals such as the *International Journal of Cognitive Ergonomics*.

The memory remembers if it can write a story and build a link between scenes. The memory remembers if it can find the narrative link between what moves it and the product it is supposed to remember. The product you are placing to be remembered must fit into what the actor is experiencing. Its use must be relevant to the plot.

The audience does not remember the brand of the drink that the actor sips absent-mindedly during a tense conversation with his arch-enemy. They will remember it if the actor, while drinking it, detects a taste that reminds him of his childhood and goes back into the past in the form of a flashback. He asks his mother what is in the drink. His aunt, seeing that his mother is embarrassed, intervenes and whispers in his ear: 'It's a secret formula that made you what you are now.' This is great product placement! The producer will show reluctance and ask for a lot of money, but will eventually agree. You are his or her bread and butter, not the audience. Pay, and the customer's brain will reward you a hundredfold.

For a product or message to be remembered, the company must integrate it into a story or make a story out of it. The product will be better remembered if it has a beginning, a middle and an end and if it involves characters, adventures, evolutions and uncertainties that are resolved in the end. This also applies to television advertisements, product packaging or campaigns. Brands as different as Aston Martin, BMW and Dom Pérignon understood this when they embarked on the adventures of James Bond. What story does your offer tell? Your customer must be able to tell it in three sentences.

Pleasure and memory: the same area of the brain in action

Improve the customer's memory via pleasure. Hedonism is important in marketing, in particular in the luxury industry. Pleasure increases the memory and encourages purchases. But what pleases the customer? We tackled physical pleasure earlier on.

We have sensory organs, holes in our skin surface through which you can see the inside of our body. To enhance their sensation, these holes are not protected by a thick layer, our skin. Through these holes – tongue, eardrum, retina, and genital mucosa – our neurons have direct access to stimulation and friction. Our neurons are arranged on these organs in such a way that regular friction produces more electricity and more sensation than elsewhere. Pleasure seekers scan a painting with one look, let the friction of music reach their eardrums, or let the chef prepare food that will excite their tongue.

Companies have understood that, in order to get customers to buy, they must be rubbed in the right place and at the right pace, through playing with contrast, which is the stimulator par excellence, at the right pace: too slow and the excitement threshold is not exceeded; too fast and customers get irritated. There needs to be contrast on the retina, on the eardrum, and so on.

This technique for physical pleasure also applies to mental pleasure. The contrast and flow of information sent to the customer's brain by the communication campaign, the new product and the selling method are likely to create mental pleasure designed to predispose the customer to buy. These are no longer physical contrasts (colours, shapes, luminosity, etc) but mental contrasts (meaning, values, goals, etc). For memorization through pleasure, we refer the reader to the Stage 2 recommendations.

Always impact the three types of memory

To increase the customer's memory, simultaneous entries must be used. The long-term memory is made up of three elements:

- Your history, stories and personal experience are stored in the episodic memory by scene or date.
- The data, facts and encyclopedia are stored in the semantic memory by category.
- The procedures, gestures and know-how are stored in the procedural memory by objective.

This results in the following neurological advice:

- *Associate words with actions.* To increase the memorization of your product, get customers to perform a special, unusual (but not difficult) gesture to grab, hold, open or use the product. They will remember it all the more, as the brain equates it with know-how. What is the 'movement' of your product?

- *Categorize your offer.* Your product must be easily classified by the customer into three increasingly small boxes. In which box does the customer think your offer fits? Furniture/cheap/trendy?

- *Stage your offer.* Your product must be an event, including dates, circumstances, etc. In winter? By the fireplace? With friends?

- *You only have the time granted to you by your brain.* Increase the customer's memory via sequenced entries. Firstly, capture the sensory memory. You have one second to dazzle with your poster, to open the short-term memory. After this, make an impression in the short-term memory. You have one minute to convince that you will give pleasure, that your product fulfils fundamental needs. You have one minute to make an appointment with the long-term memory. After this, delve into the long-term memories. You have one hour to present your offer, to share the experience, to engrave the data and to get customers to perform the actions. Don't forget the booster shots in the form of seasonal campaigns, one day, one week and three months later.

Becoming unforgettable is also remembering your customers

How do you increase your own memory? By improving your self-organization

To increase your own memory, there are few proven methods that are easy to implement apart from the following three:

- Read out loud the important sentences and figures to be memorized. They enter via two cerebral channels, which confirm each other.

- Stop reading for one minute every 10 minutes, to give your brain time to put things in order.

- Change the type of mental task every hour, to give your neurons time to recharge.

Add an appendage to your short-term memory

The other name of the short-term memory is evocative: 'working memory'. It retains, for a limited period, only what is necessary for the next task. A little person writes the question asked and the decision to be made on a whiteboard and, for one minute, runs up and down the corridors of the long-term memory in search of answers. The little person is thinking.

The intelligence of an individual is assessed by the capacity of his or her short-term memory. The bigger it is, the longer the little person in search of answers to the question will have to bring back answers and the better their quality.

A person equipped with a short-term memory with a two-minute capacity has greater intelligence than a person with only a one-minute capacity. The longer we have to decide, the better the quality of the answer is, up to a point. To put it simply, the short-term memory of animals is one second and that of humans one minute.

A team of individuals can constitute a far bigger short-term memory by constructing a collective short-term memory, in the form of a decision table. The problem is clearly written down in the form of a question that requires a specific answer or under which each team member will write down the pros and cons.

Buy a whiteboard and put it in a prominent place. Write down the important decisions that you need to make in the form of one or several simple questions with a yes or no answer. Below the question, draw two columns: arguments in support of the 'yes', arguments in support of the 'no'. Leave it for a day or two. Progressively add elements to the columns as the ideas come to you. After several days, the answer becomes obvious. You have created an appendage to the team's short-term memory. Its capacity is one or two days instead of one minute. You have gained time between the question and the answer. You have become more intelligent.

Memory is everywhere in the brain, in the form of neurons, in many areas. There are no dedicated memory areas, but if some specific regions of the brain are damaged they seriously handicap the memory.

From a functional perspective, and to put it very simply, customers have a long-term memory where they store their past experience, and a short-term memory that allows them to decide whether or not they want to buy the offer. To be blunt, and scientists will excuse us, the short-term memory can contain at most one minute of information, after which it must be emptied by making a decision, by emitting a prejudice. This first impression that you have made upon your memory will taint and influence the entire course of its reasoning. 'The first impression is correct.'

You have entered the memory. You must now activate it. You must extract and get customers to remember at the right moment so that they can buy. This is the fifth stage of the Neuromarketing method. Be beyond suspicion: force a subconscious decision.

Be beyond suspicion

10

Satisfy the customer's subconscious – Stage 5 of the Neuromarketing method

You have successfully passed the fourth stage. You have entered the customer's memory while increasing your own memory. This is the best position for getting the customer to decide, subconsciously initially and then consciously. The customer must sign up. To ensure this, you must:

- increase your leadership;
- play with the customer's instincts and shortcuts;
- work on the customer's mimesis.

Why is this important? You must become accepted by the customer's subconscious, because if it blocks your product there will be no sale.

You can be more easily accepted by the subconscious. Its decisions are predictable.

Influence the customer by increasing the leadership of the product and salesperson

Enhance your physical appearance to impress the customer

The customer's brain, which is not very familiar with the salesperson, must decide within seconds whether or not this person is dependable or trustworthy. Rapidly forming a first impression is a matter of survival. Friend or foe?

Eye-tracking studies show that the customer's brain subconsciously examines the salesperson in search of elements that will provide proof, or not, of his or her quality. These studies reveal that the subconscious spends a lot of time examining the salesperson's watch and shoes. The brain implies, for no logical reason, that if the shoes and watch are quality products the person wearing them must be dependable and listened to.

Studies show that we subconsciously place greater value and authority in tall, thin and flat-stomached people. If you show photos without presenting any other arguments, most interviewees decide that people with this type of physique have more value and authority than others. Despite the fact that these physical characteristics are not connected in any way to our value as human beings, our subconscious brain expresses its opinion.

The salesperson's physical appearance, however, has no influence on customers if customers take the trouble to listen to their conscious brain instead, but it is extremely slow, and this salesperson is asking for a quick decision!

Be on your best behaviour when dealing with the customer

- *Eat slowly and small amounts.* During business lunches, professional salespeople eat small amounts, slowly and spend little time choosing from the menu. The customer's brain will perceive this attitude as 'dominant'. This can only be the attitude of someone used to being obeyed.

- *Do not complain, and do not justify yourself.* A complaining sales representative gives the impression of not being in control or responsible. How can a customer follow a complainer? Sales representatives who constantly justify themselves give the same message to the customer's brain, a message of uncertainty, when the customer needs certainty at the moment of the purchase!

- *Always wear a sign of authority.* High-quality clothes, pens or titles facilitate the sale. The title of vice-president was invented in companies for sales representatives. A tidy desk is a classic sign of authority, a sign of control and leadership.

- *Oblige the customer.* Giving away something free at the beginning of the relationship obliges customers. The notion of exchange is imprinted on their subconscious brain. They will give something back, and will be more likely to buy.

Manage availability and accessibility

- *Increase visibility, and reduce accessibility.* For the customer's brain, visible is important and rare is expensive. Is it contradictory to ask a salesperson to be less accessible? On the contrary. This difficult access gives the customer's brain the message that the salesperson is an important person who must be followed. However, reducing salespeople's accessibility does not mean reducing their availability. Well-organized sales representatives always follow up customers by calling them back, making sure they get them the information from somebody else, etc.

- *Increase similarity, and reduce familiarity.* You must initially open the door to the customer. The customer's brain will only open up if the salesperson appears familiar and similar to him- or herself. The rule 'Different is dangerous' still applies. When customers half-open the door, wipe your feet on the mat and their reflex will be to open. However, once the door is open, the salesperson must lead the way, and therefore maintain a certain distance from the customer, avoiding being too familiar.

- *Never show any sign of stress.* When purchasing, the customer is stressed and looks for certainty. The sales representative must represent this certainty. Never show any sign of weakness in front of the customer.

Influence the customer by playing on the brain's shortcuts

Truism and mimesis promoted by the customer's stress

The world is too complex for our brain. It can only survive by simplifying things. To cope with this, it uses a number of shortcuts, as described in Chapter 2, which reduce its intelligence but increase its efficiency.

These shortcuts become powerful decision-making aids commonly used in marketing approaches. They sometimes rely on truisms, ie unverified truths that become credible because they are repeated so much. Rumour creates a perception of truth: 'If you say good things about yourself, it gets repeated and everybody ends up believing it.' Rumours are sometimes persistent, such as the poor quality of Japanese cars or the hidden esotericism in certain logos such as Procter & Gamble's.

As also described in Chapter 2, the brain's natural tendency towards mimesis constitutes a key element leading to the customer's decision to purchase. Imitating purchases, or buying what others are buying, is driven by the customer's stress. In times of crisis, the best thing to do is choose the same solution as others rather than be creative.

Current research on the customer's subconscious decision

Will your offer satisfy the subconscious of your customer?

Answering certain questions requires the help of a knowledge engineer:

- How could your offer be modified, including the salesperson and the virtual or physical retail site, so that its leadership, instinctual nature or mimesis is increased?

- What symbolizes your product's strengths, has positive connotations and is easy to understand, even if this has no connection with it? An animal, a country of origin, a specific shape, a fruit, a character or a standard type of behaviour?

- How could your offer trigger a better imitation reflex in your target customer?

To answer these questions, knowledge engineers help you assess how your offer is evaluated by the 24 rules of automatic intelligence, and how it can create shortcuts in the brain.

Current research on the customer's subconscious decisions

Our brain contains a conscious and a subconscious part. When we talk to someone, we think about it. When we drive a car, we don't think about it.

If it happens in the cortex, it is conscious and voluntary. If it happens below, it is involuntary and subconscious. Listening is a conscious action. Hearing is subconscious. This description is somewhat simplistic, but easier to understand at first.

We do a lot of things subconsciously by applying old information-processing rules validated by life experience.

Some rules are conscious ('I compare prices'), while others are subconscious ('Rare is expensive'). The human brain decides in its own special way. When faced with the unknown, it must make a quick decision to protect itself, to determine if there is danger and to establish whether it must attack, run away or stand still. It must make a decision in less than one minute, and decide if what is coming is good or bad.

When its time is limited (eg an unknown situation or a scenario moving too fast), the brain uses rapid but low-quality decision-making rules. The brain makes decisions on vital issues in less than one minute. Marketing has one minute to convince, all the more so as the brain always regards the first pieces of information received as the most important.

Marketing below the radar

Numerous studies have shown that good-quality subliminal advertising is efficient under certain increasingly well-known conditions.

Asking employees to smile is subliminal advertising. Customers purchase a little more without making a conscious connection between their action and the employees' smiles.

Using positive words in an advertisement is subliminal advertising. This adds nothing to the rational description of the product but directly targets the customer's subconscious.

Associating your brand with objects of desire is subliminal advertising.

Conveying a purchasing message without really revealing who you are is important for unglamorous products: tobacco, banks, fuel, etc, which must conceal their true identity and sell an atmosphere of pleasure conducive to their consumption.

The customer's subconscious is giving you the all-clear. Its decision is positive: it buys. However, customers are not just animals. They are very intelligent, and sometimes rational. They have signed up and chosen your offer, but they can return it. They may not purchase again from your stores. They leave your store and come to their senses. Your ultimate weapon for total satisfaction is to be irreproachable. This will be tackled in the sixth and final stage of the Neuromarketing method.

Be irreproachable 11
Satisfy the customer's conscience – Stage 6 of the Neuromarketing method

The subconscious has given you the green light. The final stage is not the easiest. Now you need to satisfy the customer's reason:

- Help the customer make the right decision. Correct the biases of the customer's decision.

- Offer the customer what really suits him or her. To do this, you need accurate customer segmentation.

You must satisfy the customer's reason. Customers are not just animals. They are equipped with genuine intelligence, which calculates a quality-to-price ratio and does not just accept the packaging. If let loose, this high intelligence can prevent a purchase. It must be negotiated with. It can put aside emotions, instincts, immediate needs and even sensations.

There are two ways to satisfy the customer's reason. This high rational intelligence is increasingly well known by knowledge engineers and ergonomists. You can help customers make the right decision in their own best interest, via an ergonomic presentation, by helping them correct their decision biases, instincts, illusions and emotions. Or you can adapt to customers' intelligence by manufacturing products actually designed for them, because efficient customer relationship management (CRM) and accurate segmentation have improved your knowledge of them.

Help the customer make the right decision

We buy and then we regret it. Our brain is constantly making small mistakes. Marketing sometimes exploits these mistakes in order to sell something. The legislator protects customers, for example, by giving them a reflection period or by making sure salespeople do not forget to provide them with information.

Marketing can also play the game. It can help customers avoid their natural decision biases by giving them better information.

A presentation of the offer that stimulates the intelligence

When faced with a choice, the brain asks itself four successive questions:

1 What is the best way to achieve my objectives that never exposes me to any danger?

2 What are the obstacles to my objectives?

3 How do I make the best use of my resources to overcome these obstacles?

4 What series of actions must I undertake to adapt to these obstacles and achieve my objectives using a limited amount of my resources?

The offer must spontaneously help the customer's brain to answer these four questions:

1 Does this promise, or this offer, expose me to any danger or does it help me achieve my objectives?

2 What are the obstacles (price, time, access) inherent in this promise?

3 How can I minimize using the resources I possess to overcome these obstacles?

4 How can I adapt and, consequently, what should I do to benefit from this promise?

All corporate brochures and websites spontaneously answer these key customer questions, in the right order. Verify your own. Your customers will have a better understanding of your offer.

You must not fear customers' rational intelligence. If you have a quality product and if you do not promise more than you can deliver, it will not hinder the commercial relationship.

If you provide customers with the correct information, you will be irreproachable.

Respect speed limitations

The customer's brain is too slow to take in everything you send it. In a rapidly evolving situation, it receives up to five times more information than it can process. This explains why we often feel as if we have too much work, why we can never finish our day, or why we have the impression we are writing a draft of our life without ever being able to finish it.

These complaints are irrelevant from the moment we accept having a brain and a certain freedom. Those who know the brain well have learned not to complain about having too much work. This would be like complaining about having to breathe. They have learned to be happy and a little disappointed, and not to search for perfection or to try to understand everything.

Marketing messages should be calibrated to fill the pipes leading to the brain at the right pace and in the right amount. The norm is approximately 800 words equivalent per minute: any fewer, and customers go elsewhere; any more, and they get stressed.

Upgrading a poster, packaging or presentation to these 'brain' standards requires a little work, but it is worth it. Customers will be grateful. Knowledge engineers will help you achieve this.

Give customers time to think

Give them time

If the brain has less than three seconds between the question and the answer, it bypasses the highest form of intelligence; it does not have time. This is why we give better answers in writing than in a conversation.

Let them sleep on it

'Sleep on it!' as your grandmother used to say. She was right. The brain receives too much information during the day, which it classifies at night. It shelves the information left behind during its paradoxical sleep. The following day, you have access to more information received the day before.

Try this with your mail. Do not send it straight away. Wait until tomorrow to read it over. You will be surprised at how much you will improve it.

You will be irreproachable if you let customers think, change their mind and leave the store without buying anything. You will lose a purchase in the short term but you will gain loyalty in the long term.

Do not get them to make impulse purchases

This is what you find in books teaching you how to make better decisions:

- 'Never sign in front of the salesperson.'
- 'Shopping is cheaper if you have a list.'
- 'Do not change your decision in a meeting.'
- 'Before buying, leave the store for at least 10 minutes.'
- 'A poor environment can lead to a poor decision.'

Why? Because the customer's brain is quickly overwhelmed by its environment, and because marketing has become extremely good at designing an environment – store, website, sales interview – making the customer a little less intelligent or a little stupid.

From a medical perspective, the more blood there is in customers' frontal lobes, the more the decision will be in their own interest. If the environment removes blood from the frontal lobes, the brain is at its mercy. It can remove blood by calling it to other parts of the brain, eg using voices, images it presents, and decor. The more active the sensory areas of the customer's brain are, and therefore the more they require blood, the less blood will be supplied to the brain's highest decision-making centres and the less they will be able to focus rationally on their purchase.

The layout of the store and the salesperson's pitch light up areas of the brain that will distract them from their own long-term interest, which is only alerted when there is blood in the frontal lobe.

More than 30 per cent of impulse purchases are avoided when customers apply the strict rule of leaving the store for a few minutes before paying. These few minutes away from the marketing environment help return the blood to the most intelligent part of the customer's brain, which kicks in once again.

How many words have you spoken and how many decisions have you made during meetings that you regretted once you returned to the peace of your office?

Do not encourage customers to make impulse purchases. Once they cool down, they will regret it and they will bear a grudge against you. You will sell once, not twice.

Ask customers to think about it. Examine with them whether they really need it. Do not get them into debt. You will gain loyal customers.

Help them calculate

Marketing, in an effort to increase prices without customers noticing, sometimes imperceptibly reduces the size while keeping the price stable.

And it works. The customer is deceived. Most customers do not consciously notice. The brain does not like calculating and is fairly bad at it; it is a poor statistician and a mediocre mental calculator. Marketing plays with the price-to-volume ratio of the product, as customers often miscalculate the price per volume, unless the retailer or legislator does the calculations for them.

Do not wait until it gets to this stage. Tell them the truth.

Be frank about the risks

The customer's brain often miscalculates the risk. For example, it generally tends to fear the frequency of a risk more than its impact.

In the insurance sector, customers often buy excessive coverage against small and frequent risks and insufficient coverage against major infrequent risks, which a computer making rational decisions would not do.

One tour operator won over many customers by adding clear information on the risks inherent in its offers, hotels, countries and so on.

Customers make a rough calculation using the references in their memory

Exact calculation and rough calculation are two extremely different circuits in the brain. When customers are making a purchase, rough calculation is activated, as it is better suited to this situation. Customers do not make exact calculations. They compare with familiar quantities. Marketing can increase the price without the customer noticing or feeling upset, as long as the price remains lower than the value estimated by the customer in relation to a familiar product close in time or space.

Some retailers are surprised when a limited increase or decrease in price radically alters the quantities sold, when significant increases or decreases sometimes make no difference to the volumes sold.

Customers making a purchase use a cerebral calculation system totally different from that used by retailers when setting prices. Hence there is a certain amount of misunderstanding and a slump in sales. Neuromarketing may help correct this by providing retailers with information on how their customer's brain calculates in their stores.

Let customers maintain their judgement

High intelligence is extremely sensitive to alcohol. A single glass of wine is enough to alter the finest judgements.

Stress makes people stupid. A little too much uncertainty, and we apply rules that provide certainty but are of poor quality (eg 'Different is dangerous').

Hypoglycaemia makes people nervous. When we are hungry, just before lunch, tests show that our fine judgement is impaired. The search for food subconsciously distracts us from our primary purposes. Never go shopping on an empty stomach or you will buy too much.

How much choice should I propose in my store?

A major retailer asked ergonomists specializing in marketing how many types of coffee it should present to its customers to give them the impression of freedom: it wanted enough choice without paralysing them with too much choice. The answer was six products, according to ergonomic standards and the study of the decision time for this product.

Excess choice is harmful. More than six versions of a product, and the customers become frustrated. If there is too little choice, they go elsewhere; if there is too much, they cannot decide.

A lot of choices require too much effort to decide and too many regrets after the decision. Research by Nobel Prize winner Daniel Kahneman (2011) showed that opportunity cost and loss of satisfaction have more impact than gains. Offering a maximum number of choices does not encourage consumption. Too much choice prevents decision and action.

Offer customers what really suits them

CRM: an essential tool for improving knowledge of the customer

Thanks to more affordable IT techniques and data storage tools, businesses have access to a wealth of information on their customers. The CRM system helps organize this information based on complex statistical treatment tools. With these tools, marketing not only can enhance its knowledge of customers but also can foresee and anticipate their needs, sometimes before they are even aware of these needs. Thanks to loyalty cards, supermarkets can identify the surname, first name, address and other information concerning the customer buying nappies for his or her one-year-old baby. Based on this information, it is easy for the computer to anticipate all the purchases the customer will have to make as the child grows older. Well-calculated promotions will offer the customer bargains by purchasing a product just before or at the exact moment he or she needs it.

The same applies to car purchases. Many customers have a herd mentality. By studying the past behaviour of certain customers, it is possible to anticipate when a vehicle will be replaced. The CRM of the consumer credit institution can propose a credit offer at the right moment, even before the customer starts inquiring about purchasing a car, and long before he or she asks the bank for a loan.

By using the information and activating the data-processing and customer relationship software contained in the CRM's 'back office' and 'front office', marketing can give customers what they expect. One-to-one marketing or mass customization can even be implemented, making it possible to provide each customer with a comprehensive marketing mix adapted to his or her needs and level of profitability and the risk the customer represents for the company (eg risk of insolvency or late payment).

The development of CRM systems is experiencing significant growth in companies that manufacture consumer goods and services with a large number of customers. CRM is a major marketing tool for increasing customer loyalty.

Use accurate segmentation

Customer segmentation has always been an essential marketing tool for enhancing knowledge of the customer. In addition to traditional segmentation criteria such as age, gender, habitat, income, profession, social level and wealth, psychologists and sociologists have attempted to define new segmentation criteria. These criteria are largely based on lifestyle and sociocultural behaviour. Neuroscience proposes other criteria, more adapted to how the brain works. Neuromarketing proposes a segmentation according to new criteria, including:

- level of intelligence;
- intelligence age;
- emotional intelligence;

- resistance to stress;
- personality.

At the end of this final stage in the Neuromarketing method, provided everything has gone smoothly, you should be 'neurocompatible'.

Certain modern marketing approaches are, often unknowingly, also 'neurocompatible'. In the final part of this book (Part IV), we will develop some of those that can be used for more in-depth exploration of the Neuromarketing method.

12 Neuromarketing in application

From cognitive optimization of product conception and display to sales and communication

Companies frequently make different types of requests of Neuro-marketing. Sales departments often ask for sales representatives to be trained in selling based on knowledge of the brain. It requires more efficient marketing techniques. Marketing and communication departments want cerebral imaging studies, cognitive surveys and standard customer intelligence tests. They also want to improve existing practices in the following domains:

- more accurate cognitive segmentation to improve customer satisfaction;
- cognitive optimization of products;
- cognitive optimization of sales outlets;
- cognitive optimization of posters, packaging, brands, etc.

Publishers and film and show producers sometimes use 'script doctoring' services. They request that the audience or readers be kept in suspense, on the edge of their seat. They ask for the correction of errors in the information flow that could make the film or book boring and unsaleable.

Neuromarketing in businesses

The Neuromarketing study: some application experiments

To respond to companies' requests, Neuromarketing knowledge has been concentrated and 'packaged' within a specific programme of studies. The objective of these studies is to understand how to increase company revenue by designing ergonomic products and sales, ie optimized in relation to how the customer's brain processes the information.

Over the past 10 years, many companies have been involved in these studies. A few recent examples are as follows:

- A company in Burbank decided to make the scenario of one of its most famous films more ergonomic, to optimize constraints and enhance viewer pleasure. The studio requested that the flow of information sent to viewers' brains be regulated so as to keep them on the edge of their seat despite their familiarity with the story. Various possibilities were tested by guinea pig viewers equipped with electrodes (heart rate, breathing pattern, sweating, etc). The application of the rules of the 'right' plot was controlled by a cognitive ergonomics programme. And it works! Have you seen *Titanic*? Were you moved? And yet you knew the story.

- A New York publisher wanted to make a future book more ergonomic. The writer agreed to go along with this. He let the publisher use a cognitive ergonomics programme to test his manuscript. Simple things were tackled first, such as suppression of overly long sentences and pointless repetitions, followed by level two corrections: information flows, controlled contradictions and ergonomic editing. Finally, level three corrections were undertaken, based on test readers' pupillary reactions, and EEG waves compared with eyeball movements. The book is a success.

- A German management software company wanted to make its flagship product more ergonomic, an application designed by programmers while managers make more emotional decisions. This required translating the data into information, and information into useful knowledge. The true story of the management cockpit emanated from this approach.

Other examples of requests by companies include the following:

- A Swiss bank in Zurich wanted to make its customer relations more ergonomic as part of its marketing campaigns.

- A British bank in London wanted to make its front-of-house customer service more ergonomic.

- A Paris company wanted to make its customer texts and messages more ergonomic.

- A communication company in Paris wanted to upgrade the ergonomics of its messages to optimize images, shapes, etc.

- A Franco-German company wanted to upgrade the ergonomics of its aircraft cockpits to minimize pilot error.

- A private bank wanted to revamp its website to adapt it to the decision-making methods of wealthy individuals' intelligence.

- A perfume company wanted to implement ergonomically designed computer-aided perfume production.

- An insurance company focused on reorganizing the range of products in its portfolio in accordance with the very specific way the human brain assesses risk.

- A Brussels company conducted three 'Top 10 Sales' Neuromarketing studies (see '"Top 10 Sales" programme' below) to ascertain whether training salespeople could significantly increase sales.

- A US software company wanted to use cognitive ergonomics to design its future interface with users.

- A company in Paris wanted a Neuromarketing study to optimize its hotel receptions.

- A company in Lille, a chain of fashion stores, wanted a Neuromarketing study to optimize sales.

- A consumer goods company wanted to create ergonomic range extensions for its flagship products.

Use of the Neuromarketing method by the company's marketing function

The offer benefits from audits from various perspectives, in particular during the design phase. The Neuromarketing method integrates six traditional approaches of a marketing audit:

1 The company's standard *customers*. Who are they really? What is their neurological identity? What is their intelligence?

2 The *product*. Is its existence a necessity? What is its existence in relation to what the brain needs? Are its design and interface ergonomic?

3 Do the *poster*, packaging and campaign comply with ergonomic standards?

4 How is the *brand* managed in relation to senses, memory and instincts?

5 Are the *offer*, range, store and reception efficient?

6 Is the *marketing* efficient?

There are two types of Neuromarketing programmes most sought after by the company's marketing function.

Customer segmentation

Natural buying potential inventory (NBPI) tests bring segmentation closer to how market intelligence works. NBPI tests make it possible to adapt the offer to the specific characteristics of priority targets. They include intelligence quotient (IQ), emotional quotient (EQ) and stress resistance tests.

Design of new products or adaptation of existing products

These include sales increase programmes via the cognitive optimization of the product, and the marketing cockpit relating to product offers:

- *Sales outlet organization:* self-selling stores; 'Sales Point' programme designed to increase sales via the cognitive optimization of physical or virtual sales outlets;
- *Brand, poster and packaging optimization:* an 'emotive' offer; 'Story' programme aimed at increasing sales via an emotive presentation;
- *Coaching courses for sales representatives:* highly intelligent salespeople; 'Top 10 Sales' programme aimed at increasing sales by increasing salespeople's intelligence;
- *Marketing studies:* design of specific sales aid software; design of in-depth questionnaires for standard customers; studies of eyes, faces and secretions; cerebral imaging studies.

Description of the major kinds of Neuromarketing programme provided by consultancy firms

'Sales Point' programme

The 'Sales Point' programme aimed at increasing sales via the cognitive optimization of physical or virtual sales outlets (20 days for a chain of stores) consists of:

- optimizing sound and smell;
- organizing the customer's interface with product 'facing';
- store–salesperson coordination;
- website ergonomics;
- website–store combination as part of a multi-channel approach;
- displays;
- organizing surfaces and volumes (floors and walls);
- organizing entry, traffic flow and exit;
- product 'facing', messages and information;
- shape and location of the furniture;
- salespeople's behaviour in the sales outlet.

'Story' programme

The 'Story' programme aimed at increasing sales via an emotive presentation and plot puts the emphasis on:

- characters;
- settings;
- stress;
- joy;
- fears.

'Top 10 Sales' programme

The 'Top 10 Sales' programme (in two days) includes answering the following questions:

- What happens in the customer's brain when you sell?
- What happens in salespeople's brains when they sell?
- How do you impact the customer's memory so that he or she can remember?
- How do you dress?
- What mental exercises should salespeople do to prepare themselves?
- How do you influence the customer's decision?

Ergo-marketing programme

The Ergo-marketing programme (in 20 days) helps adapt and improve the presentation of the product to its customers (display, design, packaging, instructions, display panel, product visual, interfaces, 'look and feel', etc):

- Increase customer pleasure and satisfaction:
  - Stimulate the customer's aesthetic sense.
  - Stimulate the customer's hedonism.
  - Stimulate the customer's dominance.

- Increase customer productivity:
  - Stimulate the customer's senses.
  - Stimulate the customer's memory.
  - Stimulate the customer's intelligence.
  - Stimulate the customer's action.
- Increase customer security:
  - Stimulate the customer's stress.
  - Stimulate the customer's health.
  - Stimulate the customer's well-being.

Neuromarketing audit

The rules of the Neuromarketing audit

Cognitive ergonomists have established lists of rules to be respected when designing new products with a view to optimizing marketing campaigns. These rules test the product or campaign in relation to the requirements of the customer's brain. They follow the neurological stages involved in the purchase:

- capturing the attention;
- appropriate perception and sensory congruence;
- pleasure;
- positive classification memory and recollection;
- emotions;
- the decision-making process;
- preparation for the purchase;
- taking action.

A website must comply with certain rules to attract attention and be well understood. A poster campaign will be rendered more ergonomic by studying the written text.

A Neuromarketing cognitive ergonomics audit can be conducted at the design stage of the product or campaign, in which case it is called *design ergonomics*. It can be conducted at a later stage, in which case it is called *correction ergonomics*.

Experts in Neuromarketing have a variety of different names depending on their work: 'ergonomic designer' when they improve a product, 'script doctor' when they improve a book or film, etc.

Experts in Neuromarketing work for 20 days and submit a series of recommendations to make the product more marketable, because it is more compatible with the customer's vision, memory or pleasure.

The audit's response to Neuromarketing questions

The responses to the following questions...

- Is the object built to attract attention and be understood?
- Is the text understandable?
- Is the customer's mental workload optimized?
- Is the customer's stress optimized?
- Is the customer's memory well activated, with a positive connotation?
- Is the intended marketing message adapted to the customer's intelligence?
- Are the customer's decision-making mechanisms well controlled?

... have consequences in the following domains:

- an advertisement or a poster;
- a TV commercial;
- a selling method;
- a new product project;
- a package;
- an instruction manual;
- a display or a store;
- a website, a blog or a viral video;
- a brochure.

To please human brains and be bought, the product must be designed and sold in a manner compatible with how human intelligence functions.

The 'Ergos' of the Neuromarketing audit

Below are the different Ergo-marketing programmes available:

- Ergo 9001 to optimize written communication;
- Ergo 9002 to optimize images and shapes;
- Ergo 9003 to optimize plots and stories;
- Ergo 9004 to optimize environmental constraints (office or sales outlet);
- Ergo 9005 to optimize memorability;
- Ergo 9006 to optimize the potential pleasure and trust conveyed by an object or environment;
- Ergo 9007 to optimize the rational decision-making process;
- Ergo 9008 to optimize the imitability of a message.

The programmes are delivered in the form of two-day training sessions or up to 20 days of analysis and recommendations. Cerebral imaging and analysis equipment can be hired by the day. The experts who deliver these programmes are physicians specializing in human intelligence, cognitive ergonomists and experts in management and Neuromarketing.

Three true stories

The automatic art production case

Neuromarketing and its application in the artistic domain

A Neuromarketing company from Los Angeles receives a phone call about a film.

'What is the problem with your film?'

'Everybody knows the story by heart: it is the story of the *Titanic*; there is no possible suspense at the end. I need your script doctoring programme to keep the audience on the edge of their seats for two hours. Name your price.'

One week later, the Neuromarketing company upgrades the ergonomics of the script. The idea is to give the right flow of information to the viewers' brains to give them pleasure and stress up to the S point, ie not too much or too little, so that they are dying to know what happens next. Their uncertainty must be controlled by showing them what they need to see at the right moment.

Assisted art production has always existed. Publishers, critics or circles have always gravitated around the artists and influenced them to help them find their audience. Proust's publisher made a few alterations to the manuscript. Nobody is complaining about it. The publisher helps the author convey the intended message. This started with the spelling, and errors in terms of sounds or colours, after which the script doctor moved on to correcting structures and finally adding extra ideas. Script doctoring has risen from craftsmanship to programme stage. A specialist reviews all errors in terms of taste and communication.

Script doctoring is widespread in literature and films, but also in music and painting. While it can kill great works of art, it improves the quality of numerous minor works. A minor work of art is 50 per cent creativity and 50 per cent work and automated rules.

The script doctor lists a large number of cognitive ergonomics rules for each type of art. These rules may include:

- Killing the character who has just shown his fiancée's photograph within two minutes. Tear jerking: +5 points. Tintin jumps! To be continued. Suspense points: +17.

- Painters' nudes catch the eye. Sex sells. Provocation is appealing.

- The very format of a painting (size, materials, seriation, etc), depending on the era, can increase or decrease its valuation, even among the most qualified artistic experts, and outside any creative consideration.

- Giving the viewer a two-second head start on the actor guarantees tension.

While art and culture marketing already exists, art and culture Neuromarketing is in its infancy. Wanting to widen its audience is no longer a sin for a museum. Wanting to find an audience is no longer a sin for an artist. The principles involved include the following:

- *Functional magnetic resonance imaging (fMRI) can see what you think of a work of art.* If the images of your brain show that your medial prefrontal region lights up when you look at the painting, it is highly likely that, if asked whether you liked this painting when coming out of the machine, your answer would be yes. It should be noted that this area does not light up for certain paintings that you do not spontaneously like. It is possible to light up this area by presenting it to you differently and by explaining it to you, to make you like it and buy it.

- *What moves you increases the humidity in your eyes.* If you are moved by a work of art, your lachrymal glands will react and start producing before you have time to say anything. They are detected by the device, even if you are not crying. This is a good sign for the artist.

- *Computers that paint.* Apart from works of genius, which are in a class of their own, minor works of art still classed as quality by experts and curators meet certain common criteria. Despite being the result of the artist's application of ready-made 'recipes', a lot of these works have been deemed worthy of museums. The tricks of the trade are listed, inventoried and automated. Computers paint via a program that has analysed the 1,000 top emerging works of art on the website ArtMarket. What are the most pleasing and moving colours, shapes, symbols or subjects? The most advanced programs add a random function to make it look even more authentic. It is possible to submit the work of art for a Turing test. If a human jury is wrong half of the time about whether this work of art was produced by a computer or an artist, the test is positive for the machine.

- *There are several fields of research for art Neuromarketing.* Neurophysiology studies the systems of the brain that process the information received from a work of art. In cognitive ergonomics, the ergonomic recommendations made to the artists take into account how the viewer's brain perceives their work.

Appeal to a wider audience

A book, a film or a work of art wishing to widen its audience can adapt to the brains that will read it, watch it or listen to it. If the purpose of the work of art is to give the brain food for thought, move it, please it, be remembered by it or capture its attention, cognitive ergonomics can help the designer follow rules that promote acceptance by a wider audience.

This research is also being conducted by expert ergonomists, who verify the application of writing rules and give the brain food for thought. In the films *Titanic* and *Avatar* and in the book *The Da Vinci Code*, these rules are obvious, whether they were applied intuitively by the author or recommended by a script doctor.

Reader or viewer brain manipulation rules include:

- Playing with the viewer's memory by dividing the storyline.

- Never ending a story without providing elements of the next one.

- Playing with the viewer's stress. The viewer's uncertainty should score 200 points.

- Managing the rest of the information given to the viewer. The viewer acquires a piece of information before the actor or character. The viewer already knows what the character does not.

- Playing with the viewer's decision-making process.

- The plot must get the viewer to make a decision every 10 minutes.

There are 47 more rules, well known by art ergonomists.

The biology of passions

Art and culture as consumer products have common characteristics that interest neurologists:

- *Symbolism.* How does the brain process symbolic information?
- *Hedonism.* The involvement of the limbic lobe and dopamine in pleasure.
- *Aesthetics.* How does the brain process information relating to shapes?
- *Emotions.* How does the brain manage emotions?
- *Multi-sensory aspects.* How does the brain integrate the information conveyed by its various senses?

The Selling System for the Financial Consultant (SSFC) case

Before

Bank teller: Have you seen all our products? We should be able to find something suitable for you.

Customer (in a hurry): I'm a little pressed for time. We'll discuss it another day.

After

Bank teller: A red Visa card, that's what you need.

Customer (in a hurry): I had thought about it. I see that you have it ready for me. Can I take it?

What is the difference? SSFC.

The first conversation between the bank teller and the customer has too often been overheard in this major bank. It was predictable. The bank teller is young and rushed off his feet. He is an administrative agent, not a salesperson. The contact with the customer lasts two minutes. There are far too many financial products for him to remember in such a short time. How do we train this overloaded employee in selling? How do we help him when he does not know how to sell? The Neuromarketing method has been applied to solve this problem and increase sales.

A decision-making aid system for tellers, mimicking the human memory algorithm and avoiding its decision bias, was developed in collaboration with two major retail banks. A simplified expert system presents to the teller the three products that the customer is statistically most likely to purchase from a possible 247. They are selected according to the customer's profile, products the customer already owns, and selling circumstances at the counter such as the limited contact time.

The teller inserts the customer's card into the computer. The customer's record immediately appears, with the three products that the customer is statistically most likely to purchase, in light of his or her history and profile, clearly visible at the top.

Certain banks are equipped with an optical system projecting the names of these products on to the counter window, so that they are within the teller's field of vision without being visible to the customer, like a teleprompter for TV presenters.

The teller keeps his eyes on the customer. Failure to do so is likely to result in the customer leaving.

The 'Sales Point' case, or how to optimize sales in outlets

This is about bringing customers in, retaining them, and transforming the entry into a purchase. A sales outlet benefits from being adapted to the brain or from improved ergonomics.

The 'Sales Point' Neuromarketing method has been developed as a result of numerous surveys in stores, hotels, banks and websites, in collaboration with ACCOR, Promod, Crédit Suisse and other entities.

The brain of the customer is influenced by its environment during the purchase. A cognitively ergonomic environment, ie satisfactory for the brain, increases the transformation ratio. An ergonomic window display increases the entry ratio. An ergonomic restaurant menu increases the average bill paid by the customer at the end of the meal.

Test customers wearing eye detectors, a secretion detector, and heartbeat and respiratory rate counters browse all possible and imaginable configurations of virtual and actual stores. All their reactions are recorded.

Virtual retail stores on the internet are very useful for studies. They can be modified at will and very quickly as soon as the customers enter. Facing, sounds, geography and prices are automatically adapted to the customers' profile as soon as they are detected.

A major consumer organization that has identified this type of store explains to the customers: 'You will buy more there than elsewhere. Before you buy, leave for a few minutes and come back only if it is to confirm your purchase decision. These stores are organized by ergonomists specializing in the decision-making process.'

Neuromarketing in application

13

Sensory marketing in the sales outlet

In light of the emergence of the internet and the increasing sales transfers from physical to virtual outlets, retailers are undergoing major changes. From simple, sometimes cold, storage areas, they are becoming welcoming living spaces. Their concept is being revamped: in addition to selling products and services, they offer play and sporting areas; they are turning into classrooms, workshops or recreation areas; they are becoming genuine urban leisure centres, free for customers and their families. They are attempting to acquire a soul that legitimizes and encourages consumer visits. Their desire is to increase customers' well-being and provide them with a pleasant shopping experience, conducive to additional purchases. Physical stores use an asset that the internet does not have: the ability to work on each of the customers' five senses. This is the mission of sensory marketing. By stimulating the consumers' senses, it strives to enhance pleasure, differentiation and the shopping experience. Its objective is to win over the customer's brain by satisfying his or her senses.

The advent of the internet has rendered the transformation of physical outlets inevitable

The increase in the power of the internet is inexorable

The internet has completely reshaped our lives. Europe has become the world's second-largest geographical region in terms of number of users. In Europe, France ranks second behind the UK, ahead of Germany.

This development of the internet, regardless of the country considered, is the combined result of an increase in supply and demand. The number of households that have access to the internet continues to rise. This growth is accompanied by an increase in the number of commercial transactions carried out by each internet user. At the same time, there is a plethora of constantly evolving websites.

The competitive relationship between the internet and physical outlets is different in each sector

In some sectors, such as the record or software industry, the internet is becoming the grim reaper for physical stores. In others, online and 'offline' stores tend to coexist and develop within the same companies. To prevent this development from favouring online activity, physical retailers must rethink and enhance their 'promise to the consumer' compared with that proposed on the internet.

Faced with the internet, physical outlets are redefining their mission

The internet has undeniable competitive advantages that cannot be ignored by physical retailers. Compared with the stores' fixed and limited opening hours during the day and week, the web is extremely flexible. It remains open 24/7 for the convenience of each user. There is no longer any need for cars, buses or trains; gone is the long wait at the checkout and the irritation that goes with it. Last but not least, the internet is generally less costly and enables realistic comparisons between the desired products or services.

Because of these advantages, physical outlets must be different. They must look for and showcase their own competitive advantages: the human touch and physical contact with the product. Unlike the internet, stores are becoming living and meeting spaces. The five senses of the consumer can be solicited to increase the pleasure of the visitor experience. Stores are becoming a place of sensory experimentation, making it possible to increase the human brain's well-being, thereby stimulating the customer's desire to stay longer and feel more comfortable in the outlet.

The longer customers stay in an outlet, the more likely they are to buy

A growing number of marketing or environmental psychology studies attest to the positive influence of a store's atmosphere on consumers' reactions. Sensory marketing is often used for the purposes of raising the brand's profile or maximizing customer satisfaction. As pointed out by Rieunier *et al* (2009), 'it would be wrong to assume that sensory marketing can only deal with the image or contribute to the shopping experience; it can also be integrated into the objective of influencing the customer's purchasing behaviour. For example, all butchers know that they sell far more chickens when they turn on their rotisserie.'

By pleasing the senses, sensory marketing helps increase visiting time in the outlet, thereby promoting the conditions necessary for additional purchases.

Importance of the senses in the brain's decision to purchase

All five senses have direct, unfiltered access to the brain

The lobes of the brain, located in the cerebral hemispheres, are the receptacles of the centres responsible for the senses. The occipital lobe, located at the back of the hemispheres, near the occipital bone of the cranium, contains the centres responsible for sight. The information relating to touch is transmitted to the parietal lobe, located in the middle part of the brain. The hearing and taste centres are 'hosted' in the temporal lobe. The sense of smell has a direct route to the limbic system, the centre of emotions and the emotional memory. The five senses have direct access to the brain and, according to Antonio Damasio's hypothesis, contribute to the individual's decision-making process via the emotions they convey. This is the 'somatic markers' theory.

The senses awaken the 'somatic markers'

In a now-famous book, Damasio (1994) advocates the hypothesis that emotions influence the behaviour and decision-making process. This hypothesis is based on the in-depth study of human cases, including those of Phineas Gage and Elliot. Each case presents the same lesion in a specific area of the front of the brain, the ventromedial prefrontal cortex. All these patients have retained normal intelligence while losing all ability to experience emotions. Based on their reasoning alone, they have become incapable of making rational decisions and learning from their mistakes. For Damasio, these clinical cases support the hypothesis of decision-making aids he calls 'somatic markers', ie all significant past experiences recorded in our brain. Often linked to the activation of a sense, they are reactivated by a similar solicitation of this sense and help rule out choice, which, based on experience, can be detrimental. This automatic process can be triggered unconsciously (in which case it is referred to as 'intuition') or by consciously activating an emotion. Unlike Descartes, who advocated a clear dichotomy between mind and body, rationality and emotion, Damasio believes that emotions, often reactivated by the senses as illustrated by Proust's madeleine, contribute to the speed and pertinence of reasoning. By diffusing the smell of cedarwood in their stores, is French brand Nature et Découvertes not attempting to evoke our childhood, as this odour is strongly reminiscent of that of a freshly sharpened pencil?

New organization of sales outlets to appeal more directly to human intelligence

The organization of stores, whether category killers, which are more accustomed to it, or mass merchandisers such as hypermarkets or superstores, is being progressively reviewed. The objective is to respond to a consumption logic that appeals to the perceptions of the human brain. To respond to the logic used by the human brain to influence the writing of a shopping list, the products are arranged in departments that correspond with the 'customer's reality' and not with the logistical reasons that previously prevailed. The breakfast department, previously spread throughout the entire store, now groups together all products relating to this theme. The baby department groups together food, nappies and all equipment necessary for the baby's well-being in the one place. To improve the consumers' awareness of this new organization and ensure they do not waste their time or become irritated, retailers emphasize the dividing lines between departments. Clearly visible signs are attached over the departments or in aisles. Different-coloured furniture is used, as are distinct floor coverings: carpet in the toy department to increase cosiness, floorboards in the wine or clothing departments to promote the perception of a quality image, and linoleum in the food departments of 'discount' areas to reinforce the low-cost image.

Diffusing specific fragrances for each universe helps the human brain to decipher this organization, such as the baby powder smell in the childcare department or sun lotion in Bloomingdale's swimsuit department. The use of music or sounds specific to the universe (eg knife sounds at the meat counter, seagull cries at the fish counter) helps consumers identify it.

Stores deliberately organize the transmission of information intended for the five senses of the customers. While sight has always been the focus of attention, smell, hearing, touch and taste are widening the range of sensory marketing in sales outlets.

Multi-sensory experience

Each sense has a specific influence on the customer's brain. Their harmonious combination as part of sensory strategy and positioning makes this influence even more effective. Sensory marketing focuses on creating a 'congruence of the senses', a genuine 'Neuromarketing mix' of all five senses (Hultén, Broweus and van Dijk, 2009; Krishna, 2009).

Smell

Smell is an essential sense for humans, having made a significant contribution to the survival of the human species by enabling the detection and avoidance of the enemy or rotten food. It plays a major role in the memory. The sensory receptors of the nose are directly connected to the limbic area of the brain, which hosts the memory and emotions. Smell affects a substantial part of our daily emotions, which themselves are involved in our purchasing decisions.

Butchers and bakers are well aware of the positive influence of the smell of roasting (fat and thyme) or freshly baked bread, which they have been using for a long time to attract customers and increase sales. In addition to the positive emotions they reactivate, certain scents trigger, in the brain, the secretion of gastric juices, which leads to the search for food. This physiological reaction produced by certain scents is increasingly used by retail chains. This takes on even greater importance when food samples are offered, which makes consumers feel indebted. As a result, their likelihood to buy increases.

The use of scents in stores is facilitated by technological progress. The nebulization technique encapsulates a fragrance in a paper or textile support. This constitutes a starting point for the invention of scented wrapping ribbon or branded scarves. As underlined by Rieunier *et al* (2009), this technique is proposed by Hollywood Chewing Gum to retailers to diffuse the scent of mint via automated devices every time someone approaches them. This has led to a 10 to 25 per cent increase in sales. Similarly, a major amusement park in the Paris region saw its sales of popcorn increase by 20 per cent after diffusing this scent within 200 metres of the kiosk where this food was sold.

Certain scents such as lavender, sandalwood, rose, orange and vanilla aim at relaxing customers, while others such as jasmine, chamomile, lemon and mint favour stimulation.

Olfactory signatures are being developed in several areas, such as the hospitality sector. Having conducted experimental tests in three hotels, the Air Berger company has been appointed to develop a new olfactory signature called 'Cosy Lounge' for the lobbies of all Novotel establishments. Test results show that the olfactory atmosphere has a positive and significant influence on the chain's modern image and on the overall level of satisfaction and recommendation. It seems that an olfactory signature policy has also been determined for Mercure hotels, another chain of the Accor group, and Le Méridien hotels of the Starwood group.

Many companies are positioning themselves on the olfactory market, such as Air Aroma (Australia), Firmenich (Switzerland) and Givaudan (France), in an effort to create olfactory identities for brands in accordance with marketing briefs.

Touch

Lindstrom (2010) points out that 50 per cent of purchasing decisions are made without a reason. Rieunier *et al* (2009) claim that 75 per cent of these decisions are made in the sales outlet and approximately 80 per cent of the products handled are bought. This would explain why a large number of retail chains are now displaying costly products on shelves, with protection, when they used to be presented behind protected display cases.

To be purchased, these products must possess certain characteristics that can be successfully deciphered by the brain. For example, the brain computes the 'right' weight a product should have to be perceived as a quality product. If the product does not fit this criterion, it is likely to be put back on the shelf and, as a result, not be purchased. Neuromarketing studies are particularly enlightening for designers in this domain. A well-known mobile phone manufacturer, following Neuromarketing studies, decided to artificially increase the weight of its devices. Significant sales results have been observed.

The importance of touch is not limited to weight. A hotel chain decided to standardize the texture of its sheets and blankets with a view to creating an identifiable tactile sensorial element for the brand.

The use of touch is all the more important when female customers are targeted. Women are known to have 10 times more tactile sensors directly connected to the brain than men. For women, and men to a lesser extent, touch also triggers the secretion of two hormones, oxytocin and prolactin, linked to affection, attachment and a feeling of tender love: hence the pleasant effect felt during caresses or even just massages.

Certain clothing brands, in particular female ones, focus on the texture of their products to satisfy the brain's sensory perceptions associated with touch. As a result, their sales have increased significantly, in particular when touch is combined with smell. The Antoine et Lili brand invited its customers on a sensory journey based on touch combined with smell in its textile stores. Princesse Tam Tam, a female lingerie brand, is reorganizing the texture of its fitting rooms to provide its customers with a pleasant sensory place.

Taste

Taste releases chemical substances in the brain. Certain scientists claim that those released by chocolate can partially imitate effects similar to those of marijuana by affecting the chemical substances of the brain, triggering a euphoric reaction.

Tasting a product during a hungry period as part of a point-of-sale promotion forces the brain to produce gastric juices, which increase the desire to eat. As consumers are not rude, they do not finish what is on the tray but buy the product instead. They do so all the more willingly because the brain, having received something, has a natural tendency to want to give something in return.

Taste studies are not limited to the distribution of food or the catering industry. In the sport sector, the Decathlon company studies what taste its snorkels should have to satisfy the brain of swimmers.

Hearing

For the brain, silence is reminiscent of death. A minute of silence is observed after the death of leading personalities. Surveys conducted among customers and salespeople in stores show that these two categories prefer a musical atmosphere rather than silence, unless the music is overly repetitive or obsessive for the personnel, like Christmas songs for example. In the United States, the Musak company provides retail chains with functional music that is supposed to increase sales and the productivity of the sales staff. Sixième Son designs audio logos specifically created for retail chains. Brands like Célio, Starbucks and Nespresso have ordered their own music.

Certain restaurant chains have noticed that customers tend to order more drinks, wine, coffees or desserts and stay at the table longer when the background music is slow.

The Atoo Média agency offers a multi-channel audio approach to retail companies, which includes reception, waiting and sales areas, the telephone (eg switchboard, voice servers, and music on hold), and websites and blogs (eg games, banners and animations): all events and media that can be enhanced by sound.

In addition to sales areas, Neuromarketing studies relating to sound are undertaken for many products available on the shelves. In Europe, the Adriant company works on the sound of crisps in the mouth that is right for the brain; Eurosyn works on that suitable for a hairdryer. In the United States, NeuroFocus also conducts studies on the brain's perception of sounds.

Sight

Nearly a third of the human brain is devoted to sight (Eagleman, 2011). Sticks and cones convert the light into nerve signals transmitted to the brain. The signals detected by the eye are analysed by the brain, which compares them with past experiences. During a hunt, it calculates the size and speed of the game and compares the new images with the memories of similar forms, colours and movements stored in its memory. The same applies in a sales outlet. The forms, colours and arrangements of products on shelves have a direct influence on the customer's brain. NeuroFocus in the United States, with its Mynd device, a portable electroencephalogram (EEG) that transmits data to a computer in real time via Bluetooth, conducts numerous studies designed to optimize the consumers' view in sales outlets. Knowledge engineers who use analyses derived from Neuromarketing confirm certain observations such as: consumers examine more items when the light in the store is stronger; neon lighting gives a bargain range a cheap impression; and a professionally enhanced restaurant menu significantly increases the consumption of the dishes selected.

The clothes of the salesperson or adviser influence the perceived quality of a product or competence of a service: the chef's hats worn by the sales staff of the famous Paul bakeries or the red overalls worn by sales advisers in Bricorama DIY stores. The ambiance on the shelves is essential for giving retail brands a sensory connotation that pleases the brain.

In Asia, particularly in China, Carrefour has abandoned the cold colours of its European sales outlets in favour of warmer colours. Red and yellow, which symbolize prosperity and happiness, are favoured to give a feeling of festivity in the brains of Asian consumers.

In Europe, the Histoire d'Or jewellery brand exudes a sense of luxury and festivity by creating a design made of warm colours and open spaces for its stores. Delitraiteur, a grocery store and caterer, opted for wooden shelves to create a cosy atmosphere. In the highly professional domain of pharmacies, the Dragon Rouge agency has managed to make the Viadys chain more welcoming in a sensory manner by using an original, warmer layout than that found in traditional pharmacies. Vichy, the famous mint manufacturer, is reviewing its packaging to give it the chalky aspect of its products.

Certain design agencies such as Dragon Rouge and Carré Noir believe that the appearance of sales outlets should be revamped at least every five years, which also helps increase sales by 15 to 40 per cent. These agencies sometimes suggest a less significant renovation every three years.

The choice of fashionable colours for the years to come is always complex. To avoid making too many mistakes, retailers are often assisted by the trend forecasting agencies present in most countries.

Convergence of the senses and the increased use of Neuromarketing approaches to improve sensory marketing in sales outlets

Improving sensory marketing in sales areas requires enhancing their coherence, searching for a genuine 'marketing mix' of the senses to create a sensory identity. It must correspond with the positioning chosen by the retailer for its brand, and must be perceived to be logical by the customer's brain.

Through its applications, sensory marketing in sales outlets can considerably benefit from advances in Neuromarketing.

Convergence or 'congruence' of the senses

Two conditions must be met to maximize the efficiency of sensory marketing in sales outlets: 1) Define a clear, specific and differentiating positioning to develop a pertinent sensory trademark for the customer's brain. 2) Ensure that the choices made to please the five senses are coherent. The search for a conductor capable of guaranteeing this coherence (or 'congruence' between these senses, as experts call it) is a necessity.

Certain retailers have had interesting success in this domain. In Europe, Nature et Découvertes has managed to create a form of well-being for the customer's brain closely linked to nature. It has done so by developing a congruence between the scent of cedarwood reminiscent of the sharpened pencils of our youth, natural products, easily accessible and touched by consumers, and a musical atmosphere evocative of nature.

In several countries, brands such as Abercrombie & Fitch, Sephora and Nespresso focus on coordinating selected ingredients to catch the attention of the five senses, with a view to dramatizing their stores and creating a genuine sensory experience for visitors.

The search for the convergence of the senses designed to please the consumer's brain is not exclusive to luxury goods retailers. In Belgium, the 'leader' of discount outlets, Colruyt, wishes to provide its customers with a sensory experience based on values relating to asceticism, frugality and smart shopping. The brand essentially targets consumers who have values other than consumption and who are not susceptible to excessive solicitation. To convey this impression, their sales areas are designed with undecorated flooring, neon lights, products presented in transport boxes, and refrigerated units replaced by cold rooms.

The congruence of the senses does not only relate to sensory marketing actions for the retail brand. To be fully effective, it also requires strong commitment from the personnel: clothing, origin, training, passion and attitude. The human resources department is involved from the recruitment stage.

Nike stores in the United States primarily recruit sports fanatics. In Europe, Flash 76 recruits motorcycle enthusiasts. The 64 stores in the Basque Country preferably employ devotees of this region's culture; the same applies to the 66 stores in Catalonia.

The congruence of the senses is a difficult policy to undertake for retail brands, as it often requires the hiring of actual knowledge engineers (or cogniticians) and the verification of their hypotheses by Neuromarketing studies.

Neuromarketing approaches and studies

Most self-service purchases made in sales outlets, but also those made on the internet, are spontaneous. They are guided by the customers' emotions, which to a large extent control their judgement. These emotions have often been overlooked by traditional marketing and communication approaches, which try too hard to rationalize the act of purchasing. Lindstrom (2010) claims that purchases are largely conditioned by the emotions of the consumers, who subsequently attempt to rationalize them. One of the reasons for underestimating the emotional factors in purchasing decisions is probably that traditional marketing surveys find it very difficult to measure them.

The emotions experienced by customers in stores and on the web are largely conditioned, as seen in this chapter, by the state of their five senses when they decide to buy. In traditional surveys based on interviews in all shapes and forms, people find it difficult to translate their emotions or the perception of their senses into simple words. Most of the time, consumers have an insufficient grasp of sensory terms. Unlike experts who can codify words relating to senses, for non-specialists the same word can cover different notions depending on the person who uses it. Studies can be affected by two biases. The first stems from the difficulty in choosing pertinent words for the different consumers within the same panel. The second is due to the fact that this type of analytical approach can also influence them. Furthermore, as stated by Giboreau and Body (2007), certain forms of survey are not suitable for collecting reliable information in this domain (eg reluctance to express feelings to an interviewer, or a decision not to complete the questionnaire by mail or e-mail because of the difficulties in answering the questions).

Neuromarketing research, by directly examining the brain's reactions, can be more pertinent for understanding the emotional reactions of the consumer's brain when his or her five senses are directly stimulated. The EEG used by the different Neuromarketing research firms in sales outlets or in relation to the internet can be very efficient. Other less costly studies described in previous chapters, such as those on facial analysis measured by video cameras connected to computers by wifi, laser pointers, telemetry or hormone secretion, can also be useful.

PART III: KEY POINTS

- By applying the knowledge gained from neuroscience to marketing, the Neuromarketing method makes it easier to improve the efficacy of its action. It is designed as a six-stage process progressively making it possible to win over the customer's brain and then customer loyalty. At the end of these stages, the customer is no longer a customer but a partner.

- The first stage consists of attracting the customers' attention and good will by soliciting all their senses. The customers' senses are the doors to their brain and purchase decisions. When suitably solicited, they make it possible to gain customers' attention, followed by their pleasure, and finally they access customers' memory. Neuromarketing strives to satisfy the customer's nose, ears, eyes and skin, introducing its stimuli and messages through all doors at once. Its ultimate weapon is the convergence of the senses.

- The second stage aims at making the product indispensable via the fundamental needs it fulfils. The primary objective of the brain is pleasure. At this stage, it must achieve pleasure by providing dominance and games. This is necessary to satisfy the fundamental needs of its primitive brain: sex and food. The Neuromarketing method must strive to respond to this. In this respect, it uses the stimuli favoured by the brain, highlighting sex and dosing the intelligence with the hormones that create pleasure, in particular dopamine. Human beings like to play, and the brain likes to dominate. To enhance the efficacy of the marketing action, the products must demonstrate their ability to respond to the brain's primary needs. Neuromarketing will also improve its pertinence by letting customers play and dominate.

- The third stage focuses on developing an offer designed to satisfy customers via their emotions. It helps retain customers before they consume and gets them to return afterwards. The

Neuromarketing method during this stage helps move customers during and after the consumption of the product. This is a key phase for gaining their loyalty and ensuring they move up the range. Three methods are used to achieve this, which consist of managing the customers' emotions, stimulating stress up to a point, and showing them films so that they can be transported. One of the best ways to move them is to make a film of the offer. In addition to the customer, sales representatives must also be constantly maintained at an optimal level of stress, or 'S point', to guarantee maximum efficacy.

- The fourth stage aspires to being unforgettable by profoundly affecting the customer's memory. The Neuromarketing method helps penetrate the customer's memory by using the right language, the right repetitions, the right sequences, the right stories and the right kind of pleasure. For better efficacy at this stage of the process, marketers must also use neuroscience to increase their own memory capacity.

- The purpose of the fifth stage is to ensure that the customer's subconscious suspects nothing. The idea is to reassure the customer via the brand and convince him or her to join up. To achieve this, the Neuromarketing method recommends an increase in leadership, stimulation of the instincts and customer intelligence shortcuts, and focus on their mimesis.

- The sixth stage should result in the company and its offer being perceived as irreproachable by the customer's intelligence. Up until now, the Neuromarketing method managed to sell to customers largely by satisfying their subconscious. To make sure they return and to gain their loyalty, it must now satisfy their reason. It is necessary to help them make the right decision by appealing to their intelligent brain and to help them think by correcting the bias of their decision. What they are offered must match their needs rather than respond to a primary and fleeting impulse. The use of accurate customer segmentation, based on the maturity of the intelligence, helps complete this task. Customers will be grateful for the help in making an informed choice matching their fundamental needs rather than a spontaneous purchase conditioned by an impulse decision. Their loyalty is secured if Neuromarketing manages to adapt to their expectations and develop a genuine partnership with them.

- The Neuromarketing method has concrete applications already tested by major corporations: they relate to product innovation, communication, distribution and sales. It focuses on the business world as well as films and various forms of artistic creation.

- Important applications of the Neuromarketing method can be observed in the cognitive optimization of products, sales and communication and in sales outlet sensory marketing.

PART IV
Perspectives for today... and tomorrow

Marketing is being extensively renovated at the dawn of the 21st century. This renovation has been imposed by changes in its environment:

- *How the consumer behaves, attaching greater importance to emotion over logic.* In a world where progress is accelerating at an exponential rate and technology is becoming unavoidable, the emotional quotient (EQ) is gradually outweighing the intelligence quotient (IQ).

- *The internet invading most homes.* It is becoming an essential tool in the decision-making process, sometimes a veritable drug for the brain. It paves the way for instinctive decisions, where 'zapping' replaces reasoning. It forces communication to change radically and become interactive. The traditional marketing consumer is progressively becoming a proactive consumer.

- *Saturation of the consumer's head by messages* makes the brain less and less available to receive new ones in their different forms (television, radio, press, billboards, internet, etc).

- *The multiplication of offers creates a permanent level of stress in the customers perusing the shelves of hypermarkets or the internet.* Customers are becoming like a 'chameleon in a kilt'. The security of their choices is reinforced when they find a famous and attractive brand or a retailer investing in quality to enhance their loyalty.

- *Increased sensitivity to offers emanating from socially responsible companies.* Interested in sustainable development, the environment or a cleaner planet, consumers are looking for offers proposed by socially responsible companies.

The success of tomorrow's marketing will come from its ability to understand not only the customer's tastes, needs and expectations, but also his or her intelligence and how to satisfy it. This is what the Neuromarketing method intends to achieve.

Certain recent marketing approaches are headed in this direction, most often subconsciously. When they are adopted and lead to success, it is often because they match, voluntarily or otherwise, the deepest expectations of the customer's intelligence.

In the following chapters, we will select five of these approaches that we believe correspond with the principles and recommendations of Neuromarketing:

1 value innovation to surprise the customer's brain;

2 permission and desire marketing to avoid saturation and rejection by the customer's brain;

3 interactivity to improve communication with the customer's brain;

4 brand policy to reassure the customer's brain;

5 quality to retain the customer and legitimacy to leave the customer's brain with a clear conscience.

Value innovation to surprise the customer's brain

14

Consumer goods and service companies are about to engage in a ferocious battle to gain the recognition of the customer's intelligence over the next few years. Marketers and sales representatives will be on the front line. To have the best chance to win, they can no longer afford simply to be good and satisfy the customers' expectations. Their offers must be exceptional, even surprising. Success will come from the marketing function's ability to create innovation in all marketing mix policies, capable of surprising and satisfying the customer's intelligence. To achieve this, the use of traditional marketing approaches and corporate strategy will be insufficient. The marketing of tomorrow must look for new methods such as those emanating from 'disruption' or 'value innovation', striving to discover 'blue oceans' for the companies' offers.

Although these methods have emerged recently, they are not new. They are the result of a reflection that initially emerged via the evolution of fundamental research in the United States. Having been tested by charismatic leaders of companies recognized for their success in business, they constitute new approaches in terms of strategy and marketing. Their common feature is that they surprise the customer's intelligence without trying to be better than, but by being different from, the competition.

Traditional marketing makes extensive use of strategic matrices to prepare its decisions. Porter, McKinsey, AD Little, BCG, Ansoff and other matrices are found in numerous plans emanating from this discipline. The theories of competitive analysis highlighted by Michael

Porter and the benchmarking strategy proposed by RC Camp are frequently used.

To prepare its strategic reflection, innovation marketing tends to refer to theories based on the search for unexplored regions, breakthroughs and differentiation. Since 1995, US authors such as Hamel, Prahalad and Moore have stressed the obligation to create new markets, and no longer simply define strategies by taking the competition into account. Handy (2002) describes our era as 'a time when the only prediction that will hold true is that no predictions will hold true. Change used to be more of the same, only better. Those days are gone. Now change is discontinuity.' Thomas (1995) believes that 'the level of discontinuity is measured by how a company learns to deviate from the stated requirements of the market'.

De Bono (1996) presents 'lateral thinking' as a way to wander off the beaten path. All these authors agree on encouraging companies to forsake traditional thinking, adopt discontinuity, and adhere to what Peters (1987) refers to as 'chaos management': surprising and challenging the customers' intelligence.

Developing an innovation marketing policy relies on discontinuity theories more than on traditional strategy and marketing theories. Our purpose is not to carry out an exegesis of the exhaustive literature in this domain. To prepare an innovation strategy, we have selected, for illustration purposes, the reflections of three authors, Dru, Kim and Mauborgne, whose research has been implemented in the form of models in communication agencies for the first and consulting firms for the second and third. Careful reading of their research is essential for any manager wishing to put value innovation into practice in his or her company.

Disruption for improved communication with the customer's intelligence

Disruption to create change

Creating highly desirable products and services for the customers' intelligence and proposing a highly emotional relationship involve significant risk taking. This is all the more necessary as the competition is watching and heavily investing in marketing and communication. Taking risks involves straying off the beaten track and going against preconceptions.

These preconceptions include the desire to bow to the consumer's expectations, the fear of cannibalizing one's own products, and the fear of no longer continuing to do the same thing just because this has been successful for many years. As stated by Dru (1997), Barry Diller, CEO of Fox Television, believes that most companies are still 'slaves to market research': 'If you ask consumers what you should do, the answers are mundane or conventional. Unconditional respect for consumers hides a lack of inspiration and, more seriously, an excuse for conservatism.'

Mobile phones would never have been invented in certain European countries if telecom companies had followed the results of market research, which often revealed their utter pointlessness in the eyes of the consumers interviewed.

D'Aveni (1994) stigmatizes those who believe they can indefinitely hold on to existing advantages. Peters (1987) mentions Steve Ross, who claims that 'staff should be fired for not making mistakes'. To efficiently implement innovation marketing, it is necessary to consider change as an ally and to create the market by being more daring. Dru, in his 1997 book entitled *Disruption: Overturning Conventions and Shaking Up the Marketplace*, proposes a three-phase method or 'discipline' – convention, disruption, vision – making it possible to present a breakthrough strategy. Reading this book can be beneficial to any marketer opting for the development of a new, innovative and different form of communication, likely to surprise and then convince the customer's intelligence.

The disruption method

In his book, Dru (1997) defines disruption as 'a way of thinking which defies established convention and creates a new vision for companies, capable of making their brands grow quicker'. The idea is to develop a breakthrough strategy, making it possible to propose exceptional products and brands conveying new notions and allowing customers to dream. Disruption 'is a non linear, discontinuous discipline with three theoretical stages: convention, disruption, vision'.

Convention

The idea is initially to detect the conventions of a market, the body of preconceived ideas that maintain the status quo. Deviating from conventions will give rise to the idea of disruption. Conventions are all the things that our intelligence blindly accepts, as they are so anchored within the company. People do not even think of challenging them, because they are so widespread. They can be of a marketing nature, emanate from the customer, come from consumption, and so on.

The initial idea is to establish the list of conventions, in particular 'ready-made' opinions, to measure the gap between these opinions and reality. The staff must be trained, using suitable techniques, to clear their minds of accumulated habits and knowledge. Dru organizes, in his communication agencies, 'convention meetings' during which there is a brainstorming on preconceived ideas within a company, an industrial sector, an area of activities or a market. Weeding out conventions helps distinguish between certain immutable ideas, which are difficult to modify, and others that are not.

In 1984, conventional wisdom was that computers were reserved for an elite of specialists. Not so long ago, it was common practice to believe that the main purpose of skin care creams was to make users look younger or, in the publishing sector, that a book had to be launched in the afternoon. In the retail sector, marketing departments long believed that communication must be based on tangible elements, that replenishment is crucial in the clothes selling business, that it is impossible to obtain lower prices than in hypermarkets' central purchasing units, and so on. Refusal to bow to these conventions has allowed Apple, Oil of Olay, Harry Potter, Virgin, Zara and Lidl to achieve success against competitors that have remained conventional. Differentiation that creates surprise attracts the attention of human intelligence. The resulting strategy based on a business model and original positioning must obtain the approval of the executive committee while satisfying the consumer's intelligence.

Disruption

Disruption consists of challenging conventions to determine whether they continue to be effective or whether they are maintained because it is easier or simply out of habit. The marketing department attempts to find disruptive ideas that go against convention. Apple challenged the point of view that computers were designed only for IT professionals by making the Macintosh widely accessible to a larger audience. Oil of Olay in the United States tackled 'youngsters' by selling 'beauty at any age'. Virgin prefers to give its stores an emotional dimension by targeting youth culture rather than proposing tangible offers. Zara refuses to carry out any stock replenishment. Lidl is structured to obtain lower prices than those of hypermarkets' central purchasing units. Conventions are tenacious because they facilitate the decision-making process ('We must proceed as usual, and do what we are good at' or 'We must do what others do') and reduce the stress of novelty.

Disruption, by breaking from convention, makes it possible to travel to unexplored territories. It encourages the company to place itself in what Kim and Mauborgne (2005) refer to as the 'blue ocean'. The purpose of disruption is not so much to invent as to rediscover certain fundamental ideas buried by habit, by taking a step back, broadening the scope of observation and improving insight. Disruption is all the more effective as it targets areas where companies have a 'herd mentality', where established brands are losing momentum. It is useful when the company needs to create highly desirable products and services for consumers or proactive consumers who are no longer satisfied with standard products or the routine management of their relationships with the company. The customer's brain can be won over and reconquered only by ceaselessly providing it with attractive and constantly renewed products, services, relationships and distribution methods, offers that appeal to its intelligence. The purpose of disruption is to help companies and their brands make 'strategic leaps' by defying convention, promoting creative disorder and stimulating change.

Vision

Doing the opposite of what others do is not enough. Believing it may lead to failure. Bic, by launching a cheap perfume that challenged convention, learned this the hard way. Disruption does not consist of opposing or contradicting convention but of contributing something else, a new vision of the market and the brand. Ford, Apple, Nike and Starbucks provide this vision. Ford wants to democratize cars; Apple wants to free consumers from the tyranny of computers; Nike wants to give people the desire to pursue excellence; Starbucks wants to create coffee connoisseurs similar to wine connoisseurs.

The vision can be based on serious studies, and its feasibility can be verified. However, it is not sufficient if it is not partly built on dreams. George Bernard Shaw wrote: 'Some men see things as they are and say why. I dream of things that never were and say why not?' Disruption leads to success if it brings pleasure and a dream element to the intelligence.

To be desirable, a brand, product or service needs a vision that appeals to the customers' imagination, emotions, affections and fantasies rather than to their rational behaviour. It also needs customers to feel the sincerity and even passion of the company's designers and staff via their products, distribution channels and communication. Mike Nike's passion for sport and effort is felt through the Nike brand. The company must present a point of view in order to be heard over the cacophony of brands. MAIF in France are asserting themselves as a militant insurance company, claiming that they are proud of their ethical values and can reject profitable customers who do not share these values. Even if they do not share the company's vision, consumers are inclined to respect it. Choosing the type of customer with whom managers, marketers and personnel really want to work can become a key success factor for a company.

The value innovation approach: the 'blue ocean' strategy

Theoretical foundations of the 'blue ocean' strategy

W Chan Kim, Boston Consulting Group Henderson Chair Professor at INSEAD, and Renée Mauborgne, also a professor at INSEAD, present an interesting approach for reflecting on the development of an innovation marketing strategy. Through a wealth of research in their book *Blue Ocean Strategy* (2005), they propose a 'value innovation' method for companies working in a saturated competitive environment where all products look alike and a price war is raging.

To break free from these constraints and get out of what the authors call the 'red ocean', companies must make a 'leap in value', ie a genuine strategic shift resulting in the creation of a totally new market space, a 'blue ocean'. US and European companies cited by the authors, including The Body Shop, Cirque du Soleil, eBay and Swatch, have been surprisingly successful because of their ability to open unexplored marketing regions and create new demand. The development of innovation marketing can succeed only when underpinned by an audacious, original and innovative strategy. The reflections of Kim and Mauborgne have been adapted to the business consulting field by Marc Beauvois-Coladon, Marketing Director of Barnian Consulting, a member of the Blue Ocean Strategy Network. They provide customers with value innovation. They can be used as a basis for the creation of a pertinent business model and the search for appropriate positioning as part of the development of an innovation marketing strategy. We reproduce hereafter the broad lines of their approach, illustrated by concrete applications carried out, during a conference at HEC Paris for Microsoft executives, by Marc Beauvois-Coladon.

'The only way to beat the competition is to stop trying to beat them.' To illustrate the meaning of this sentence, Kim and Mauborgne imagine that the universe of the markets resembles two sorts of ocean. Red oceans are made up of existing companies within an area of activity: the known strategic space. Blue oceans represent all companies that do not yet exist: the unknown strategic space. In the red ocean, the rules of the competitive game are widespread and largely accepted. As the market space is increasingly congested, profitable growth prospects are dwindling. The benchmarking of close competitors is the major focus of strategic studies. The primary operational response is the search for productivity gains so as to engage in a price war. The products are becoming standardized because all they do is copy the competitors. As a result of fierce competition, 'the ocean is turning blood red'.

The search for a blue ocean is characterized by the quest for unexploited strategic spaces and the creation of new demand. Marketing looks less at neighbouring competitors and more at other sectors and markets likely to inspire a different strategic or business model. Blue oceans are uncharted territories where competition is irrelevant, as the rules of the game have yet to be defined.

For many years, research in terms of strategy has primarily focused on the study of the red oceans of competition and the neighbours' competitive advantages. In the near future, faced with significant and rapid change in the environment, including accelerated technological progress, globalization and evolving consumer behaviour, the domination of monopolies will disappear. *Businesses must reinvent most of the strategic and marketing theories that have emerged over the past century and are becoming obsolete.* To survive, they need to create new, uncontested strategic spaces. These spaces are the foundation of innovation marketing.

A study conducted by Kim and Mauborgne among 108 companies resulted in measuring the impact of the creation of 'blue oceans' on turnover and profit margins. The results reveal that 86 per cent of the launches were simple product line extensions, ie improvements within the framework of the existing market space. These activities represented 62 per cent of the turnover and 39 per cent of total profit. The remaining 14 per cent of launches were consistent with the 'blue oceans' logic. They generated 38 per cent of the turnover and 61 per cent of total profit. The results are impressive. The authors believe that they reflect the need to develop a method making it possible, based on value innovation, to create 'blue oceans'.

The 'blue ocean' method

This method is based on value innovation. It helps in making a 'leap in value', which results in taking the competition out of the game by creating a new, uncontested strategic space. Value innovation should not be confused with technological innovation. It occurs only 'when the company harmonizes its innovation efforts with its imperatives in terms of relevance, price and cost' (Kim and Mauborgne, 2005). The method illustrated below in 'The "blue ocean" strategy in application: the example of Thomas Cook' proposes a progressive process.

The strategic framework

This constitutes a diagnosis as well as an action tool. It takes a snapshot of the current state of the competition within the known strategic space. It indicates the domains in which competitors are investing, the authorized criteria where competition comes into play, eg products, services, and speed of execution, and the concrete advantages proposed by the different offers to the customer's intelligence. The horizontal axis represents the range of competition criteria and investment domains characteristic of the sector considered. The company located at the top of the ladder proposes more in a specific domain and invests more than others. A good score in terms of price indicates that its prices are high. It is possible to plot the curve of the different offers in relation to the essential criteria and to determine the strategic profiles or value curves of each brand. They represent, in a schematic form, the company's relative performance in relation to the criteria adopted by the competition in this sector. To design the strategic framework of the area of activity, the notions of rivals and customers must be put aside to focus on alternatives and non-customers. Ideas on alternatives can frequently come from what is happening in neighbouring sectors. Studying non-customers results in examining the criteria that explain why they do not buy into the company's products, services, distribution channels and brands. Why does their intelligence refuse its offers?

The 'four actions' model

To set aside the traditional choice between cost leadership and differentiation, the authors ask four questions:

- Which criteria accepted without reflection by the players of the sector should be excluded?
- Which criteria should be mitigated compared with the level deemed normal in the sector?
- Which criteria should be reinforced, far beyond the level deemed normal in the sector?
- Which criteria thus far neglected by the sector should be created?

Applying this rule to the strategic framework sheds new light on old truths, by challenging certain practices commonly used by the players of the sector, without discernment, and by focusing on the needs and perceptions of non-customers. It attempts to understand why non-customers are not customers of the brand. It helps in reflecting upon the creation of a new universe: a 'blue ocean'.

The 'exclude–mitigate–reinforce–create' matrix

The 'exclude–mitigate–reinforce–create' matrix complements the 'four actions' model. By requiring the completion of four quadrants, it results in the creation of a new value curve.

Characteristics of a good 'blue ocean' strategy

According to Kim and Mauborgne, the final strategy emanating from the 'blue ocean' method must feature three key characteristics to be fully effective:

- *Focus.* The efforts must be concentrated on a limited number of criteria emanating from the reflection based on the previously established matrix. Costly domains must be excluded to allow overinvestment in the domains selected by the focus on new priorities.
- *Divergence.* Divergence proposes moving away from an approach of purely reacting to the competition. The new

strategy is based on a unique and differentiating value curve, which is therefore difficult to reproduce. It must be unlike anything else in the area of activity the company has just left.

- *A catchy slogan.* A good blue ocean strategy must be summed up by a clear and catchy slogan. It must reflect the reality of the offers and guarantee authenticity and trust for the customer's brain.

The 'blue ocean' strategy in application: the example of Thomas Cook

In their book, Kim and Mauborgne present numerous examples of success via the application of 'blue ocean' strategies relying on value innovation. These include: Yellow Tail (the exceptional success of an Australian wine in the United States), Southwest Airlines (reinventing short-haul air transport), NetJets (offering companies shared ownership of a business jet) and Cirque du Soleil (the creation of a revolutionary form of circus with no animals, less acrobatics and more choreography, dance, harmony and themes).

For illustration purposes, we reproduce one concrete example described by Badoc and Beauvois-Coladon (2008) in an article intended for the insurance and finance sector: the case of Thomas Cook.

As a pioneer, Thomas Cook has undertaken 'blue ocean strategy' initiatives in every activity of its financial services centre: exchange services for small and medium-sized enterprises (SMEs), foreign exchange offices and international call centres for international travellers (loss of traveller's cheques, credit cards, etc), and travel and tourism activities. Thomas Cook provides SMEs with a service that cuts transfer time in half and significantly reduces the cost of the transactions made by the many financial intermediaries involved. However, its market share in the English-speaking world in which it is involved remains marginal.

The 'blue ocean strategy' begins with the creation of a team of international managers: in Europe, North America and Australia. This is a multi-functional team, which integrates representatives of IT, finance and human resources functions. Experience shows that involving 'novice' participants contributes to the process. Once the limited differentiation of the offer has been observed (see Figure 14.1), it is easy to reach the consensus that there is only one possible choice: divergence.

FIGURE 14.1 Differentiation of the offer in financial services

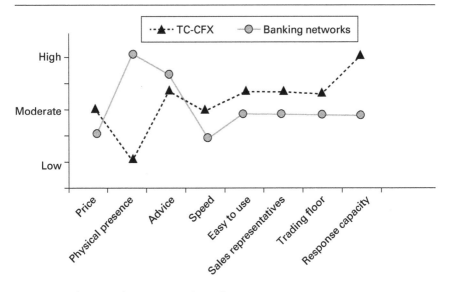

KEY: TC-CFX: Thomas Cook Corporate Foreign Exchange.

Once the experience cycle has been outlined, and the alternative options and the buyers–users–prescribers chain have been described, interview targets are allocated in a ratio of seven interviews per participant. The target of one of the participants, a financial controller, is express couriers. Each of the two teams assembled prepares four strategic profiles; the management select two, for which a pre-feasibility study is launched.

Although the idea implemented may seem simple, this is true of most 'blue ocean' ideas: it consists of informing recipients that the money has been sent and, as with express couriers, allows them to keep track of the shipment (see Figure 14.2).

Once accustomed to the service, suppliers or recipients can send the goods earlier, as they are certain of the date on which the funds will be transferred. The time it takes to deliver the goods can be cut in half.

FIGURE 14.2   Shipment by express couriers

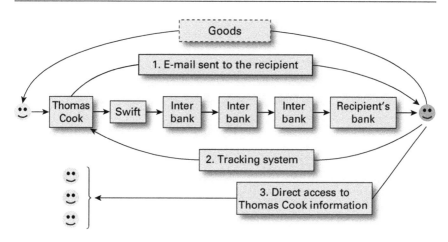

The benefit for Thomas Cook is that suppliers or recipients, satisfied with the reliability and speed of the transfer, recommend this payment method to their customers, and Thomas Cook can drastically reduce their commercial costs. Once again, exploring another area of activity helps create a divergent offer (see Figure 14.3).

FIGURE 14.3   Divergent offer created by the blue ocean strategy method for Thomas Cook

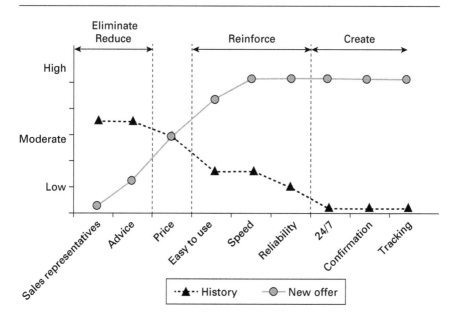

FIGURE 14.4 Results of the application of the blue ocean strategy method at Thomas Cook

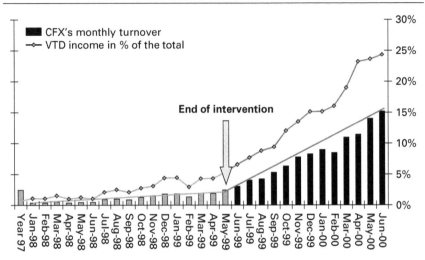

KEY: CFX: Corporate Foreign Exchange. VTD: Virtual Trading Desk.

The result was almost immediate, as the offer had convinced the customers: the activity experienced strong growth (see Figure 14.4), and the offer was one of the major bargaining chips used during the takeover by Travelex two years down the line.

Although relatively unknown in banks and insurance companies, the blue ocean strategy can become an invaluable support for their innovation policy. It will be all the more efficient, as the companies wishing to use it are pioneers in areas where the tendency is often to copy rather than stand out.

Appear exceptional to the customer's intelligence: strategies for innovation marketing

Design a pertinent strategy based on a business model and positioning aimed at differentiation

Innovation marketing strategies are difficult to develop. As pointed out by Le Nagard-Assayag and Manceau (2011), 'While successful innovations generate sales and profit margins, they remain relatively rare.' As the 3M saying goes, 'You have to kiss a lot of toads before you find your Prince Charming.'

Developing a strategy and implementing actions and an organization for innovation marketing require moving away from the traditional ideas common to this discipline and engaging in breakthrough reflections.

Initially, the idea is to call upon new thinkers and concepts striving to renovate the traditional strategic theories that were largely used in the late 20th century, as discussed in previous paragraphs. Secondly, it is important to reflect upon the development of an audacious business model and original positioning leading to a differentiating strategy that stimulates the customer's imagination and helps create attractive brands. Thirdly, actions must be implemented, largely influenced by the use of technological media, in particular the internet, and coordinated as part of an appropriate marketing mix or e-marketing mix. Finally, an organization reflecting the ambition of the challenges, involving a group of stakeholders necessary for the success of the strategy and accompanied by a change management policy, must be conceived.

An innovation marketing strategy must rely on an original and strongly differentiating business model. It represents the 'strategic trick' making it possible to win over a market segment from the competition. It stems from the previous reflections, in particular from the search for a 'disruption' or 'blue ocean'. Following acceptance by the company's executive committee, it results in the development of the positioning intended for customers by the marketing product manager.

A business model followed by pertinent positioning generally includes six features:

- *Correspond with the actual expectations of the customers' intelligence (existing or potential).* These expectations must sometimes be exposed, as they are not always clearly expressed when they relate to dreams or fantasies. They must be based on reliable studies confirming the ideas suspected by the marketing function, notably studies emanating from Neuromarketing approaches.

- *Include a distribution channel to put the offer within reach of the customers.* This is a 'b to b to c' ('business to business to consumer') approach. Without an appropriate channel, the best strategy in the world may never reach the final consumers for whom it is intended. The distribution channel sometimes constitutes the ultimate market, acting as a key factor in customer relations. This is often the case for SMEs that do not have a big enough communication budget to target a wide market.

- *Enable an acceptable return for the company on capital invested in terms of ratio and duration* (break-even point, rate of return on investment). This constraint is frequently stipulated by the executive committee or investors as part of a start-up or business creation.

- *Be sustainable in the face of the competitors' predictable reactions.* The expected evolution of the duration and intensity of the competitors' response must be dealt with professionally. Certain business models, such as the launch of active bifidus yogurts, have been genuine but short-lived successes because they underestimated the response capabilities of a competitor equipped with powerful marketing resources, which in the case of yogurts was Danone and its organic product (Activia).

- *Ensure the market adopts the proposed business model and positioning within a reasonable and anticipated timeframe.* This is a key notion, better known under the expression 'time to market'. It is particularly important for business creators keen to develop a new idea but with limited resources. The Wood-Facility Company, a start-up providing a wood wholesale 'portal', encountered difficulties because it underestimated this problem. Despite strong interest from customers, it went bankrupt after three years. Contracts were delayed when cash flow was insufficient, and the company suffered from the fact that investors who expressed an interest took too long to make up their minds.

- *Have sufficient capacity in terms of storage, production, and supply for distribution, as well as good after-sales management in the event of rapid success.* Late deliveries or poor-quality after-sales service can discourage customers from renewing their purchases from a brand perceived as unreliable. They can also enable a competitor with better organization to exploit the idea if this idea is insufficiently protected. A pertinent business model and positioning for innovation marketing cannot simply create desire; they must also structure the company so that it can constantly stimulate the customers' intelligence to renew this desire.

Innovation-based business model and positioning

The business model and positioning can create desire only if they contribute to providing the customer's intelligence with a significant innovation. At this stage, it is important not to confuse innovation with invention. An invention is largely based on the product policy. It becomes an innovation when it satisfies the customers' desire. It remains an invention when it essentially satisfies the taste of the company's internal engineers or managers.

Innovation designed to create desire is not limited to the product policy, as it concerns the customer as well as all the variables of the marketing mix (product, price, distribution, communication). The more innovative the combination of these variables is, the more likely the business model and positioning are to be sustainable over time. They become difficult for the competition to imitate.

Customer innovation

Innovation is all the more likely to be successful when it involves an issue of concern for the customer's brain in the use of a product or service (Renvoisé and Morin, 2002). Customer innovation focuses on customer segments ignored by the competition. It makes it possible to create desirable approaches conducive to success. Apple innovated by developing Macintosh computers with a human face, targeting the general public at a time when the entire profession focused almost exclusively on experts. Danish laboratory Novo Nordisk, specializing in diabetes, increased its growth rate by creating a kit for the injection of insulin, the primary purpose of which was to make patients' lives easier, when the entire profession in this domain was essentially targeting physicians. Bloomberg enjoyed great success by choosing to target financial analysts and financial information operators directly rather than IT managers. Nike did not hesitate to design sports shoes intended for the untapped market segment of babies. L'Oréal and Hermès created ethnic innovations. L'Oréal took an interest in beauty products corresponding with the physical characteristics of black skin. Hermès launched the Amazone perfume in an effort to respond to the desire of Asian, in particular Japanese, populations who use little perfume and appreciate light-fragrance perfumes.

Product innovation

The creation of new products and services is the largest source of production of desirable offers. To go beyond the stage of invention, it must correspond with an actual or anticipated market expectation. Substantial research capacity is necessary for any company wishing to implement a marketing strategy based on permission and desire. The sustainability of the business model and positioning is supported by efficient patents, the implementation of an industrial process difficult to imitate, the development of a complex marketing mix, and the support of a strong brand. There are many success stories in this domain, from Coca-Cola to McDonald's and Microsoft, noticeable in all areas of activity, eg Téfal, Airbus A320, Google, Dell, Chanel No. 5, Harry Potter, Renault (Twingo, Scenic, Logan, etc), NetJets.

Price innovation

All areas of activity characterized by substantial profit margins become vulnerable when they are affected by deregulation or try to penetrate a vast and profitable market. They are in danger of being confronted with the competition of a company that takes a significant industrial risk with a view to democratizing certain products and services by slashing prices. It can do so thanks to considerable productivity gains achieved via a cheaper manufacturing process. By democratizing its offer, it targets the desire of a large number of consumers who have dreamed about the product for a long time but could not afford it. Companies like Moulinex in the past, Bic or Buffalo Grill embarked on this adventure. The most frequent example is found in the retail industry with the emergence of less costly channels. Long-protected sectors such as the pharmaceutical, jewellery or auctioneering sectors are now in jeopardy. The development of the internet is weakening many professions that rely on traditional distribution channels imposing mark-ups that consumers deem costly in relation to the service provided (estate agents, insurance brokers, travel agents, antique dealers, etc). There are many and varied experiences in this domain. Brands such as Leclerc (drugstores, jewellery), Surcouf (computers), Antix (antiques), eBay (auctions) and Meilleurtaux (mortgages) have managed to become desirable by exploiting this market.

Some companies manage to be even cheaper than the cheapest while maintaining quality via the upstream integration of a business model. 'Hard discount' stores like Aldi, Lidl or Dia manage to compete on price with the central purchasing units of hypermarkets. They sign long-term agreements with manufacturers so that they can invest in a production unit over an extremely limited number of products, for which they guarantee a purchase volume. This investment makes it possible, thanks to the productivity gains generated, to significantly reduce costs compared with a simple negotiation while remaining profitable. As pointed out by brand expert and HEC professor Jean-Noël Kapferer, sooner or later we must put price at the heart of the innovation process. It is astonishing to see how many sectors have yet to integrate the cost culture. They are focusing on pennies while neglecting millions. From this perspective, the automotive industry has made a remarkable transition by producing cars with an increasing level of comfort, safety and eco-friendliness while putting pressure on cost price. Putting price at the heart of the innovation approach concerns not only processes: cost price can be reduced while increasing the perceived value.

Buffalo Grill, for its part, achieves substantial productivity gains that it can pass on to its prices by ensuring the rigorous management of table turnover. This price reduction policy also affects the luxury industry. Mauboussin has democratized the jewellery industry by successfully promoting less costly jewels through the use of semi-precious gems. Several years earlier, Cartier launched a luxury watch at an affordable price: its famous 'Must de Cartier'.

Distribution innovation

Innovation in terms of distribution channel alone or combined with other elements of the marketing mix helps develop pertinent business models and positioning with a view to creating highly desirable offers. Oenobiol and Roquefort Papillon have enjoyed success by abandoning large retailers and focusing exclusively on the channels of professionals: pharmacists for the former and cheese sellers for the latter.

Dell's success is largely based on the use of less costly direct channels, in particular the internet.

Zara, the flagship brand of Spanish company Inditex, managed to develop a specific business model and positioning in its clothing stores via a new concept. This concept is based on original characteristics in this industry: minimal stock rotation, organized shortage as a desirability system, frequent visits to the stores, exceptional locations, careful layout of the sales outlet, and CRM to study the latest trends and customer expectations – and all this without resorting to advertising.

Communication innovation

Less frequent than the other forms of innovation mentioned above, it has a considerable impact in terms of desirability on certain products and services. It leads to the creation of attractive brands. Names and brands such as Dim, Chanel No. 5, Ralph Lauren, Nike, Décathlon and Harry Potter owe a large part of their success to innovation as part of creative and original communications promoting breakthrough and desire. The use of sponsoring, for example Nike with Michael Jordan, Dior with Sharon Stone, and Générali with Zinédine Zidane, is an interesting way to communicate in a manner that is not disruptive for the customer. Didier Reynaud, CEO of the Affiliance company, former Marketing and Communication Director of the Générali group, managed to create a profitable, effective and segmented sponsoring system, based on events with a view to creating desire from three different marketing orientations (HEC Paris conference):

1 Win over an entire sport and be socially desirable for professionals and amateurs alike: sailing.

2 Win over a large audience, contributing to spectacularly raising the profile of an insurance group: Zinédine Zidane.

3 Win over the distribution network by allowing the company's general agents to invite a lot of customers and friends to a tasteful and quality regional event: 'Générali on Ice'.

15 Permission and desire marketing to avoid saturation and rejection by the customer's brain

The brain acts like a slow computer. It cannot process a lot of information at once. When overburdened, it secretes hormones that stress the customer.

Over the past few years, communication in all its online and offline forms has considerably increased the number of messages intended for consumers. Advertising budgets are increasing exponentially in many professions to offset the decreasing efficiency of communication.

Mailshots and telemarketing operations are multiplying despite sometimes dubious efficacy. The internet, inexpensive compared with other media, is increasingly solicited to invade our computer screen using ever more creative forms. Certain US professionals are beginning to speak out against this so-called 'interruptive' communication whose efficiency-to-cost ratio continues to decrease. They claim that it is becoming unreasonable to continue spending more for less result. To convince customers, the marketing of tomorrow has to change its focus of communication. To appeal to customers' intelligence, it can no longer simply target their needs but must take an interest in their desires. To avoid saturating and interrupting consumers, the communication must ask for their permission before sending them messages. This is how permission and desire marketing is beginning to emerge, as discussed in this chapter.

Evolution in communication and saturation of the brain receptors

Evolving communication

The dawn of the third millennium heralds a significant evolution in communication in terms of behaviour as well as the media. Everything is becoming communication; all information aims for a multitude of targets beyond the intended audience.

A minor event appearing within a limited context (a historical debate between the Pope and students of a German university, or caricatures of a prophet published in a newspaper of a small country) can inflame the planet. As a result, the brands are weakened. It is becoming impossible to control all the minor events that can occur in the multiple production and distribution centres or in subcontractors and partners. Out of the blue, these events, even if they are false, misunderstood or involuntary, can lead to extreme and uncontrollable reactions overnight, totally out of proportion with the reality and likely to cause image and financial damage.

The danger is reinforced by the soaring number of media broadcasting uncontrolled information. This information becomes credible without any verification and sometimes triggers thoughtless and unreasonable reactions on a large scale. A misinterpreted speech by the Pope results in the burning down of Orthodox and Anglican churches that have nothing to do with his authority. The ubiquity of the internet and mobile phones, which are becoming the pre-eminent media of the 21st century, further reinforces this trend.

Reason is overcome by emotion. 'Zapping' has a tendency to progressively replace logic. The prevalence of emotive factors is invading the entire communication domain to the detriment of underlying trends. Only a new emotive message can quickly take the focus off an older message. Protest against the Contrat Nouvel Emploi (New

Employment Contract) in France overshadowed, in just a few days, the perceived danger of bird flu and enabled poultry sales to pick up for no new reason.

A certain loss of sense means attaching equal importance to all information received and prioritizing that which comes from the first interlocutor to speak loudest or who primarily targets consumers' sensitivity. Directly targeting the primary functions of the brain works, even though it is deplorable.

As the message is unverified, potential mystification is likely to become the norm. In light of this constantly growing trend, marketers' practices can no longer rely on the traditional rules that have proved so successful for this discipline in the past. 'The highway code cannot be applied to planes.'

As pointed out by Christophe Lachnitt, French Microsoft's Marketing and Communication Director, during a conference at HEC Paris in 2008, 'the number of blogs, a relatively recent phenomenon, exceeds 58 million throughout the world (3.5 million in France) and is growing every year following an exponential Gaussian curve. The mobile telephony market is becoming a new Eldorado for advertisers. Communication design must adapt quickly.'

Brain saturation by communication and interruption marketing

Another consequence of this trend is that the receptors are becoming saturated by multiple exposures to the media. Yankelovich, an internationally renowned researcher, claims that ordinary US citizens are exposed to 3,000 messages every day. In this context, it is increasingly difficult for marketing managers to find available space in the brain to convey a new message to the consumers they wish to convince. Brain saturation sometimes renders sales-oriented communication unreasonable.

When they do not behave as aesthetes, to the point of collecting advertising posters, consumers claim they are increasingly bombarded by communications in their personal life. They rage against advertising ghettos that prevent them from watching their TV show, and even more so when the communication interrupts their film.

They are annoyed with the multiple phone calls they receive in the evening, offering them fake surveys. They are fed up with the pile of flyers in their letterbox despite the 'no advertising' signs, with receiving sales letters, and with having to cancel a quantity of spam every morning before they can read their e-mails. The aggressive marketing that Godin (1999, 2002), considered by some the new marketing guru in the United States, refers to as 'interruption marketing' is becoming unreasonable and counterproductive. Certain retailers reach such a level of irritation in the customers' brain that they are not far from having to issue a public apology.

More seriously, in light of the consumer's indifference, companies constantly increase communication pressure, with a less and less efficient outcome. Advertising budgets keep growing, while the return on investment is often uncertain. The efficacy of direct marketing by phone, mail or e-mail is dubious at best. Creative marketers strive to find new original media every day, bombarding customers wherever they may be.

To take but one example, the perfume industry epitomizes this abundance of communication, as stated by Le Guérer (2005):

> In 1972 all it took was 10 million francs to launch a perfume; the launch of Samsara cost $50 million less than 20 years later. These budgets have continuously increased. The launch of Calvin Klein's CK One swallowed up $70 million to $100 million in 1995. According to Eurostaf, the worldwide launch of a perfume cost 400 million to 500 million francs in 1997, and the perfume itself only accounted for 10% of this budget.

Since then, budgets have continued to increase, with often extremely uncertain results. This is a widespread phenomenon in most areas of activity. Meanwhile, as pointed out by Lindstrom in his book *Buyology* (2010) based on a vast Neuromarketing survey, the advertisement recall rates continue to decline, currently achieving excessively low scores for most campaigns.

Common sense must prevail. It is imperative to stop uncontrolled marketing and communication waste, or this fundamental concept will be denigrated and abandoned. Desire and permission marketing must emerge to respond to this imperative as well as to the need to change how the consumer's intelligence behaves. Innovation must be promoted as, more than any other variable of the marketing policy, it influences purchases and satisfies the customer's intelligence, promoting customer loyalty and sponsorship.

Permission and desire marketing

The concept of permission and desire marketing

This concept, largely inspired by US authors, in particular Seth Godin (1999, 2002), is a reaction to the limitations of traditional marketing linked to the evolving behaviour of the consumer's intelligence, the dwindling profitability of the various forms of communication and the challenges brought about by the internet and customer relationship marketing (CRM). It is based on a number of logical, simple and efficient ideas, developed throughout his books:

- There is a need to abandon interruption marketing, which is too expensive and less and less efficient.

- Only communicate with existing or potential customers who give you their permission. Find desirable 'bait' so that their intelligence gives its permission.

- To counter stiff competition, it is no longer sufficient to respond to consumers' needs; you must make them dream by becoming desirable or even essential. At this stage, product and service innovation regains its fundamental importance vis-à-vis communication.

- Constantly renew the consumers' desire. Permanently reconquer their affection to gain their loyalty.

- Manage a genuinely 'loving' relationship with your customers, who will make the company or brand a key partner and therefore difficult to compete against. This is the foundation of loyalty. The customer must become a 'partner'.

- Based on this relationship, create 'buzz marketing' or viral marketing via opinion leaders linked to communities or tribes.

The offer and image must first become appealing to the leaders, who will promote them among their community.

- Play with confidence to develop sponsorship.

These ideas can be valuable only if they result in original conquering strategies, different from what market leaders and competitors in general have to offer. They must be based on a value innovation reflection. The company should be urged to leave the 'red ocean' (Kim and Mauborgne, 2005) where it is trapped, and reach the 'blue ocean' where it will be alone in an unexplored market.

To be effective, the ideas must be supported by pertinent methodology aiming at greater coherence between permission and desire, and leading to sponsorship through viral marketing or buzz marketing.

Create desire for the offer and 'bait' to gain permission

To be properly implemented, the permission and desire strategy relies on two major variables: a desirable offer and attractive 'bait'. The offer can come from different marketing mix elements: product, price, distribution, communication. The adoption of disruption and value innovation methods helps in finding the exceptional offer, never seen before, which pleases the consumer's brain. It helps propose the 'purple cows' that Godin (2002) holds dear. They create an intense emotion in the customer, followed by a strong urge to possess the product or service presented. The consumer's intelligence is in the same situation as a collector who is offered a rare work of art that will complete the collection, which he or she has dreamed about for a long time, at an affordable price. The business model and positioning are developed to put the customer's intelligence in this emotive ambiance, which leads to desire.

The 'bait' is designed to gain permission to communicate with the customer and avoid interruption. To be attractive, it must be sufficiently interesting so that the e-mail is opened, the communication is listened to, the interlocutor on the phone is responded to, or the proposition received in the post is not thrown away. Through the internet and communities, it targets the leaders' intelligence so that they can become contaminants and recommend the company, its offers and its distribution sites to the other tribe members.

This policy is implemented using viral marketing or 'buzz marketing', also referred to by some as 'rumour marketing'.

To create desire in existing and potential customers, Gabriel Szapiro, CEO of the Saphir agency and expert in 'permission marketing' as well as web-based viral marketing, proposes a three-stage approach. These stages are applicable to create a form of communication designed to stimulate desire for the offer but also for the bait. They initially focus on *creating a plot* before *provoking the unexpected* and finally *stimulating desire*.

To be effective, the bait must be offered more than once. However, consumers and opinion leaders get tired of receiving the same bait unless it is evolving, for example when it consists of disclosing exclusive information, as is the case with Microsoft, where community leaders get information before everybody else. The use of CRM makes it possible, based on the permanent study of the needs, tastes, expectations and emotions emanating from the customers' brain, to renew the bait to continue to stimulate the desire of community leaders or, more directly, proactive consumers.

Viral marketing, a tool dedicated to permission and desire marketing

Highly developed as a result of the emergence of internet communities, viral marketing, also known as 'buzz marketing' (Rosen, 2000, 2009), is an ideal tool for propagating permission and desire marketing at a reduced cost. It is based on one of the world's oldest media: rumour.

Relevance of word of mouth

Word of mouth has long been considered one of the most effective ways to propagate an idea. Bill Bernbach, co-founder of DDB Advertising and advertising pioneer in the United States, showed the path leading to this new communication concept shortly before he died: 'You cannot sell to someone who isn't listening. Word of mouth is the best medium of all. Dullness won't sell your product but neither will irrelevant brilliance.'

Word of mouth helps fulfil this function, at a reduced cost and more efficiently. The principle is simple: rather than talking directly to the consumers and running the risk of burying your message in the advertising melee, it is better to have consumers discussing your product amongst themselves. It makes it easier to obtain permission to communicate with an interlocutor. It contributes to stimulating desire for an offer or a brand if the customers themselves are convinced of the attraction:

- *It's easy*, as consumers themselves are responsible for spreading the message.

- *It's effective*, as consumers place more trust in what other consumers tell them (particularly a friend or an expert) than in an advertisement. However, structures must be put in place to control word of mouth. The brain's tendency towards mimesis means that we follow the recommendations of people with the authority of a community leader.

- *It's cheap*, as it does not involve large advertising budgets as with interruption marketing. The consumers convey the message themselves.

The relevance of word of mouth is reinforced by the communication method of internet communities and social networks, as well as by the key role of leaders whose advice, recommendations and analyses are carefully listened to and followed. According to a Taylor Nelson Sofres survey, nearly two-thirds of internet users cite word of mouth as a regular source of information, spread by e-mail, traditional channels or forums, 'chats', 'blogs' and viral videos.

How viral marketing works

The principle of web-based viral marketing is both simple in theory to design and difficult to implement. The major difficulty is controlling the dissemination of the intended message by the brand.

Implementing viral marketing consists of:

- Identifying the communities of interest to the company who communicate among themselves.

- Spotting the leaders.

- Turning the leaders into 'contaminants'. The 'contaminants' are individuals who, within the community, benefit from the confidence arising from the esteem of the members and are willing to communicate the message to 10, 20, 100 or more people. The idea is to identify and court them, as they will add fuel to the fire and allow the 'virus' to spread. Two techniques are frequently used to win them over. The first is to pay them. When Penelope Cruz sings the praises of L'Oréal, she becomes a remunerated contaminant. Internet users posting a link to Amazon on their website are also remunerated contaminants, as they receive 10 per cent of the price of resulting purchases. The contaminants' credibility decreases if the community is aware they are remunerated. The second technique is to convince them without having to pay them. They spread the message in an effort to be helpful to the community but also to reinforce their image as opinion leaders. Microsoft feeds the contaminants with new, avant-garde and original ideas with a view to reinforcing their knowledge before any other member of their community in the domains that they are particularly fond of.

- Placing the message within a context of maximum dissemination, as part of what experts call a 'hive', the buzzing of which amplifies the communication. The 'hive' is very important for promoting the brain mimesis of a large number of community members.

- Feeding the buzz with new viral messages.

The most commonly used technical tools in viral marketing via computers, TV sets and, increasingly, smartphones are blogs and viral videos. These tools must be designed creatively; otherwise they may not be conveyed by the leaders and subsequently the internet users.

Their creativity is based on two techniques widely used in advertising: sex and humour, two of the brain's pet subjects. One of the world's most famous viral videos is the film showing Paris Hilton washing a car with extremely suggestive postures for the purposes of promoting a fast food brand!

The model developed by Badoc and Szapiro makes it possible to envisage the implementation of a viral marketing process (see Figure 15.1). It is based on a differentiation strategy approach completed by an interactive strategy. The differentiation strategy is governed by the following process: create the plot; trigger the unexpected; and provoke seduction. The interactive strategy is based on three policies: viral policy; bolstering the customer's ego; and influencing the internet user and his or her community.

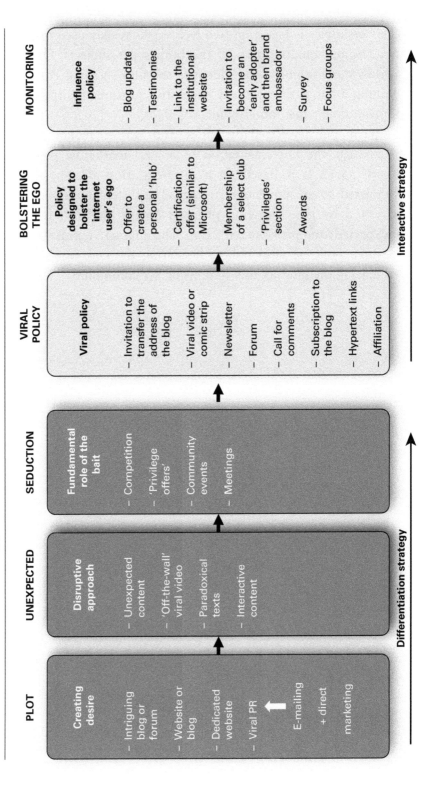

FIGURE 15.1 Viral marketing process

PLOT	UNEXPECTED	SEDUCTION	VIRAL POLICY	BOLSTERING THE EGO	MONITORING
Creating desire	**Disruptive approach**	**Fundamental role of the bait**	**Viral policy**	**Policy designed to bolster the internet user's ego**	**Influence policy**
– Intriguing blog or forum	– Unexpected content	– Competition	– Invitation to transfer the address of the blog	– Offer to create a personal 'hub'	– Blog update
– Website or blog	– 'Off-the-wall' viral video	– 'Privilege offers'	– Viral video or comic strip	– Certification offer (similar to Microsoft)	– Testimonies
– Dedicated website	– Paradoxical texts	– Community events	– Newsletter	– Membership of a select club	– Link to the institutional website
– Viral PR	– Interactive content	– Meetings	– Forum	– 'Privileges' section	– Invitation to become an 'early adopter' and then brand ambassador
E-mailing + direct marketing			– Call for comments	– Awards	– Survey
			– Subscription to the blog		– Focus groups
			– Hypertext links		
			– Affiliation		

Differentiation strategy → **Interactive strategy** →

Interactivity to improve communication with the customer's brain

16

The spectacular rise of the internet worldwide and its power of seduction are creating a new technological environment. It is frequently pointed out that, in the United States, it took the internet less than five years to win over 70 million consumers. It took cable television 25 years and the telephone 40 years to achieve the same market potential. Since 1995, considered the birth year of interactivity, the number of internet users has grown exponentially throughout the world. In 1995 for the first time the number of PCs sold in the United States exceeded that of television sets, and the number of e-mails sent exceeded that of handwritten letters. Subscription to internet services covers all countries, social classes and age categories with a few exceptions. The growth in this medium's turnover generates great optimism, even though the 'start-down' phenomenon following the 'start-up' era somewhat curbed enthusiasm for a while. There was a popular joke in Silicon Alley (New York), one of the bastions of the creation of web-related start-ups, along with Silicon Valley (Palo Alto and San Francisco): 'The e-commerce turnover will grow from $0 million in 2005 to $0 billion in 2020.' The characteristics of this medium, described in numerous publications, make it a mighty weapon for the development of products and services.

The internet owes its tremendous success to its technical functioning, in particular the possibility of proposing interactive communication, with no geographical or time limits, with the customers' intelligence. Thanks to this medium, the consumer becomes proactive. Interactivity and responsiveness are rendered efficient by the development of powerful databases combined with the customer relationship management (CRM) system, without which internet potential would be limited. The power of the internet challenges the entire marketing system, more specifically the marketing mix. It helps target the intelligence of each customer directly, providing a mass customization or one-to-one offer. Consumers tend to have less confidence in advertising when it comes to the quality of products or services; they prefer to refer to the experts of their chosen community or to the interlocutors of their social network. Marc Zuckerberg understood this phenomenon when he created Facebook, the story of which has been immortalized by David Fincher's film *The Social Network*. International social networks (eg Facebook, Twitter, LinkedIn and Viadeo) and local ones (eg Xing in Germany and Copains d'Avant in France) are enjoying exponential success.

The internet: a powerful tool driving interactivity

Long live the powerful and inexpensive web!

The success of the internet is due to the incomparable quality of this medium: its ubiquity, immediacy, economy, globalization, optimal presentation and communication capabilities. The same piece of information can be available and updated all over the world almost instantly. It can be produced by the brain of an amateur, artisan or seasoned professional and is therefore affordable to companies with limited resources.

It constitutes a global support, opening the door to international markets for a limited cost. Creative entrepreneurs can use it to promote the benefits of their products or services among interested customers throughout the world. With the world wide web, innovative solutions can quickly become known to interested customers regardless of their geographical location. The windows of the internet network can be seen everywhere and transcend national or regional borders. In addition, this permanent support is broadcast 24/7, 365 days a year, independently of time zones. It is completely adaptable to how human intelligence works. This is what has made it so successful, even though there can be significant dangers, such as addiction and other adverse effects.

The internet is a tool offering a wealth of technical communication possibilities via the use of multimedia. It enables interactivity with the customer's brain, which considerably improves the use of commercial approaches, notably based on direct marketing. It makes it possible to communicate with the virtual world, providing a large number of customers with a customized service. Finally, it is considered an inexpensive medium compared with other methods used to contact the customer. Experts estimate the cost of an e-mail at €0.20 when a letter is estimated at €1, a telephone contact at €10 and a visit to a customer at €100.

Because of all its qualities, the internet has become the tool of choice for the marketing departments of companies that wish to cultivate development strategies focusing on businesses, professionals, prescribers (eg companies that license information on businesses and corporations for use in financial decisions, or financial newspapers) and individuals.

The internet makes marketing more effective

The internet is used at different marketing levels. In particular, it helps:

- Gain the loyalty of the customers' intelligence by building regular and interactive relationships but also by rendering follow-up and customer service processes more operational.

- Increase traffic in sales outlets by bringing in new customers who have been made aware by communication and pre-sale proposals. The Fuji website is an example of how to increase traffic.

- Divert the traffic from traditional sales outlets by offering more added value and a competitive quality-to-price ratio. The competition between Dell and Compaq is played out at this level.

- Build up awareness and image at a reduced cost by targeting the brain of internet users, as is the case with companies such as Match.com, 1800flowers.com or artprice.com.

The successful development of a start-up on the internet largely depends on the number of internet users connected, on the motivation of the 'undecided' in relation to the modernism of this medium and on the purchasing power.

In the United States, web-based business-to-consumer companies have enjoyed tremendous growth. Companies such as Google, Amazon, E*Trade, eBay and Autobytel are often highlighted. French SME Laguiole made the most of this medium by successfully taking on the US market to sell its famous knives.

In Europe, internet sales are increasing exponentially. Companies such as Au Féminin, Artprice, Aquarelle and Meetic are cited as success stories. All experts agree on the fact that the future of this medium is promising. Customers are all the more interested in this medium, as it directly targets their intelligence. To be effective, it must avoid stressing or bothering them by multiplying the number of unwanted e-mails or spam. The use of permission marketing is highly recommended for companies wishing to enhance their efficiency. Its specific characteristics, such as the possibility of opening e-mails when we want to, as opposed to using the telephone, and of communicating directly with the customer as opposed to via advertisements, must be exploited. CRM also helps implement a personalized contact policy with each customer. Neuromarketing can only improve its attractiveness by allowing it to adapt to how the consumer's individual intelligence works.

CRM or how to access the customer's innermost self

The marketing power of the internet was made possible and is reinforced by the companies' newly gained ability to store and manage considerable amounts of information in confined spaces at a reduced cost. A Sony PlayStation or XBox console worth a few hundred euros can now store as much information as an entire computer room in the 1970s, which would have cost several million dollars. The CRM tool was created based on this technology. It constitutes an elaborate customer information collection and processing system that helps access the fundamental behaviour of the customer and develop a personalized and interactive relationship with his or her intelligence.

It represents a major asset for the customer retention strategy of the companies that decide to implement it. When used in an optimal manner, it helps retain customers by anticipating their needs and engaging in incomparable interpersonal relations with them. It results in the implementation of one-to-one or mass customization actions perfectly adapted to the behavioural evolutions of the intelligence's expectations, focusing on hedonism or personalization. It paves the way for sponsoring.

Thanks to CRM, the customers acquired at great cost using conquering strategies can be retained and become profitable over a long period of time. CRM is the foundation of any quality policy for each customer that the company wishes to retain. Currently implemented in companies with a multitude of customers (eg airlines, gas, electricity and telecom companies, banks, insurance companies, hotels), CRM is still in its early stages.

The difference between its technical possibilities, its desired applications and the reality perceived by the customer's intelligence can leave a lot to be desired. The relationships, often criticized by consumers, with the call centres of their telecom, electricity company or bank, based on a costly CRM system, are enough to illustrate its limited effectiveness. The problem is often that it is implemented as part of a traditional marketing approach based on the quantity of desired relationships with customers rather than quality. A bank with a large-scale CRM system, if its customers are to believe its claims in marketing conferences, rarely uses it to talk to customers about the customers' concerns and expectations. It uses it almost exclusively to incite its call centre or its sales representatives to offer customers the latest product featuring in its objectives. This product rarely matches a customer's needs, and the call often interrupts customers' activities when they have other things to do than chat politely with their banker. Contacting the telephone company for a specific reason means embarking on a hellish journey during which customers are put on hold or are in contact with staff as friendly as they are incompetent in solving the customers' specific problems, as part of a long conversation based on a fruitless search for which the customer will be billed. Nevertheless, its marketing director claims to be extremely proud of the pertinence of the CRM system during the latest international symposium. There is an enormous gap between the technical possibilities of CRM, which are considerable, and its usage efficacy, which remains limited because it does not interact enough with the customers' intelligence.

The solution can be found only if companies agree to modify their marketing approach and take into account the customers' desire by seeking their permission to communicate with them on subjects of interest to them when and where they want. Only this approach can transform this powerful relational tool into a mighty marketing weapon, after which it will overcome the competition by dissuading the customer's intelligence from going elsewhere, treating it as a genuine VIP.

The marketing battle of tomorrow, linked to significant technological investments, will be won only if a new marketing approach, focused on the new types of customer behaviour and the expectations of distribution networks, guides the technology and not the other way round. The use of Neuromarketing, at the intersection of marketing and the functioning of human intelligence, is essential for designing an effective CRM system. It must offset the errors, which too often consist of developing the system based on technical imperatives rather than human concerns. To be fully efficient and play a key support role to the commercial and marketing policy, based on the massive use of the internet, CRM must be designed by taking into account the expectations of the customer's intelligence and responding to its needs.

In addition to CRM, the development of the marketing approach via social networks now involves the notion of social relationship management (SRM).

Adapt the internet to how the customer's intelligence works

Compensate for the lack of creativity by value innovation

The innovative power of technologies derived from the internet and CRM challenges the famous marketing principle consisting of creating products according to customers' tastes, needs and expectations. The fundamental role of the customer is not called into question. However, companies taking on the internet challenge have to admit that customers are unable to express a need for a product that they are not familiar with and are sometimes incapable of imagining. They cannot help the entrepreneur with a market creation strategy. As pointed out by Rechenmann (2001), 'The customer may be king but the king is blind.' The famous quote by Akio Morita, the founder of Sony, is a perfect illustration of the new conception of the internet marketing approach: 'Markets are not designed to be studied but created.' The technologies emanating from this medium use a creative and responsive marketing approach that some authors refer to as 'creativity marketing' (Dufour, 1997), 'adventure marketing' (Rechenmann, 2001), or 'disruptive' or 'breakthrough marketing' (Dru, 1997). This challenge means that the marketing concept must evolve beyond the sense of customers, focusing on their desires and fantasies, allowing them to dream, and seeking their permission to communicate and establish an emotional relationship with them. Faced with the emergence of new technologies, the brain is overwhelmed. Owing to excessive stress when asked how they perceive the future, the customers answer conventionally when expressing their wishes. In terms of innovation, they want what they already know with some minor innovations. The customer surveys conducted by telecom companies to assess their propensity to embrace evolution towards a new telephone concept, the mobile phone, often turned out to be negative. Web technologies are evolving too rapidly for the brain of most customers. Innovations must come from methods other than market research. The use of disruption or value innovation approaches aimed at discovering 'blue oceans', as discussed in Chapter 14, seems pertinent.

Reassure the customer's brain by involving the community

To make decisions as calmly as possible, the brain needs to be reassured. Websites such as eBay and Amazon have clearly understood this expectation. They offer a number of testimonies or evaluations by customers who have already bought, which help reassure the brain of the new undecided consumer.

The development of the internet has facilitated the emergence of communities of taste and interest made up of internet users who regularly communicate with one another. The multiplication of social networks in every European country and the United States contributes to amplifying this phenomenon. Before deciding to purchase, internet users belonging to these networks consult with the community. The leaders of some communities are becoming undeniable references whose recommendations are listened to and often followed. Identifying the communities that communicate on the internet is a prerequisite for all marketers who wish to use this medium. Arousing the interest of community members requires the identification of their habits and codes for effective communication. The search for community leaders tends to prevail over the mass approach of a customer segment. Viral marketing and 'buzz marketing' based on blogs, chats, forums and viral videos are replacing traditional communication. Mass customization or one-to-one marketing will face up global approaches aimed at a specific market segment. Resorting to the community is largely favoured by the creation of the 'Web 2.0', now referred to as the 'social media'. In the United States, people are already talking about the evolution towards the Web 3.0, which is about not only connecting individuals but also linking data. The often cited slogan of *The Cluetrain Manifesto* (Levine *et al*, 2009), a ground-breaking publication in the marketing domain inspired by the social media, is: 'Markets are conversations.' Communication must now be present where the brains of proactive consumers are talking, ie in the social and community media.

Adapt the internet offer to the egocentricity of the brain: the one-to-one approach, now referred to as 'individualized marketing'

Marketing on the internet is primarily based on access to information. The recipient of the message now decides whether or not to retrieve this information. Recipients are no longer subjected to information but go in search of it. This evolution requires the implementation of a personalized relational approach, called 'one-to-one' by Peppers and Rogers (1996), the authors of two reference publications on the subject. They have also developed a personalized website: **www.1to1.com.**

CRM has evolved in parallel with the processing technologies used in the different marketing approaches. New-generation parallel computers, with exploration software, are capable of managing enormous databases at a reduced cost. They are organized in the form of rapid access matrices integrating software that can automatically build a customer behaviour model based on the analysis of the customer's past transactions as well as numerous classification characteristics. Computers, with their operating and processing software, can now expose the cybernetic intimacy of the consumer's intelligence. Following first-generation databases, focused on mass marketing, second-generation databases, far more elaborate, enable interactive, individualized marketing known as 'one-to-one marketing'. An application is presented Figure 16.1.

As pointed out by Peppers and Rogers (1996), this marketing approach is based on four key principles:

1 *Identifying customers.* The advanced segmentation of the customers and their expectations is crucial. A consumer with no expectation of a product from a company's range does not need to be solicited for actions the results of which are known in advance.

2 *Differentiating between each customer.* All customers have specific needs. They are more or less profitable for the company. Their value should determine the investment and the time that should be devoted to them.

FIGURE 16.1 'One-to-one' or 'mass customization' in a start-up company

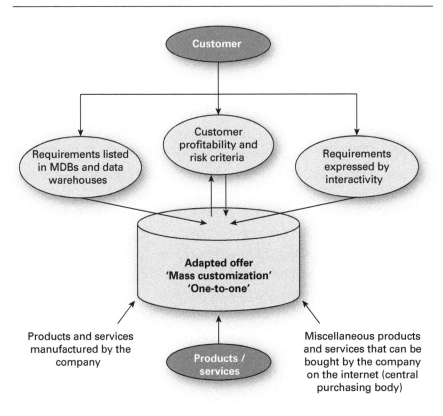

3 *Interacting with the customer.* Personalized and interactive contact is the foundation of one-to-one marketing. Every contact with customers is an opportunity to get to know them better, identify their new needs and assess their potential profitability.

4 *Selling customized products.* Producing and selling a customized product to a specific customer is the most difficult skill to implement. It allows the company to truly stand out from the competition. It is possible to personalize the offer, provided feedback from the proactive consumer is integrated into the production chain.

Implementing this type of approach, via marketing, exceeds the scope of the internet. Telephone, fax and mail are jointly used while awaiting the massive emergence of the internet on the interactive television and mobile phone market.

A new key asset for managing information is emerging, leading to the possibility of establishing personalized and permanent dialogue with the customer's intelligence and managing information in an instantaneous and continuous manner. One-to-one marketing provides customers with a response to their expectations in real time, in the form of a customized offer. This is made possible by matching the expectations and profitability level of every customer, using smart research agents, with the multiple possibilities of product and service offers. The customer's expectations are detected by means of 'data warehouses' or interactivity formulated in real time. Profitability criteria are integrated into the company's information system, which is linked to actual relationships or simulations of potential customers' emerging needs.

Figure 16.1 summarizes how to conduct a one-to-one marketing approach. The offer takes the form of a bank of gross products and services emanating from the company's manufacturing capabilities. It can also come from services developed by subcontracting partners involved in a central purchasing body that selects them from the internet. Smart research agents select products or services corresponding with the expectations expressed by the customer from the multiplicity of the offers. They shop around according to detailed specifications and return with pertinent purchase recommendations. Systems such as Bargain Finder, developed in the United States, make it possible to compare the different prices proposed for a product, select the cheapest and start the purchase.

One-to-one marketing can also adapt all commercial proposals to the value of each customer. This value is calculated provisionally via the 'lifetime value'. The idea is to evaluate what each customer brings in if he or she remains loyal for a number of years, selected according to the marketing strategy decided upon.

Adapt the internet policy via social networks to the evolving expectations of the consumer's brain

Influence of the internet and social networks on the brain

The influence of the internet on the brain grows stronger every day, to the point of creating genuine addictions, just like a drug. Since Mark Zuckerberg created Facebook, as illustrated in David Fincher's film *The Social Network*, the multiplication and evolution of social networks have reinforced this reality. Social media (formerly known as 'Web 2.0') are currently revolutionizing the behaviour of new proactive consumers. They even trigger social revolutions, overturning the dictators of the planet, as was the case recently in North Africa. Tomorrow should see the advent of the semantic 'Web 3.0'. It helps connect individuals and data. When you decide to book a theatre ticket on the internet, your agenda is automatically updated, less important appointments are postponed, an e-mail informs those concerned of the proposed changes, a restaurant is booked for after the theatre, and so on. All these evolutions and their influence on how the brain works are of interest to marketing and Neuromarketing. They must now be capable of providing companies with a method to adapt their communication to the brains of the new generations, in particular the 'Y' generation, who increasingly operate in multi-tasking mode: they can watch a film on television while responding to their e-mails and sending text messages. Their brain's attention to the different communication messages is profoundly modified. This requires significant changes to adapt these messages so that they can efficiently capture the attention of these multitasking consumers.

Traditional marketing, largely focused on conquering physical markets, is challenged by this enthusiasm for social networks. As pointed out by *The Cluetrain Manifesto*, a seminal publication for the social media, 'Markets are conversations.' The brain of frequent social media users requires that firms' communication be present where customers discuss, exchange ideas, give their opinion, or talk about brands to express their positive or negative experience. It must now track customers on their own terrain. The feelings of community leaders, of internet users who have already experienced the brand, are becoming a major decision-making factor for all targets. With the development of social networks, proactive consumers base their purchasing decisions less and less on information and more on recommendation. Research conducted in the United States tends to show that 40 to 50 per cent of internet users do not buy when community members express a negative opinion of a brand. Collective intelligence is perceived by the members of social networks as more powerful than individual intelligence. Each user becomes a producer. The crowd becomes very intelligent when it cooperates. Tomorrow's marketers will have to focus as much on conversations within communities as on traditional market segmentations.

At the same time, emotions and desire influence how the brains of these new generations respond to purchases much more than actual needs, in an environment where consumers are overwhelmed by information and a multitude of offers. Amazon's founder Jeff Bezos points out that 'Your brand is what people say about you when you are not in the room.'

Internet encirclement policy to respond to the growing influence of social networks

To respond to the growing influence of social networks on the orientation of the brain's purchasing decisions, communication policies in general and on the internet in particular must adapt rapidly. While this adaptation does not affect the efficacy of traditional advertising media, it requires a broader media context designed to respond to the interactivity expectations of proactive consumers. It means that the companies wishing to communicate effectively must gain thorough knowledge of the expectations and behaviour of their customers' brains when chatting on social networks. The use of Neuromarketing studies can largely facilitate this understanding.

Members of social networks do not want to be spoken to like customers but rather like friends. Marketers must stop acting like experts and constantly ask themselves: 'What do we talk about when we are amongst friends?' New proactive consumers are primarily interested in useful information. They want to laugh at a funny viral video, or they want to be amused by an original game. They are particularly sensitive to anything that appeals to their brain's emotions. They are particularly keen on what they can share with the community (eg a video, a game, or new and original information). In addition to their needs, they appreciate it when their desires are targeted. To comprehend this, Gabriel Szapiro, CEO of the Saphir agency, proposes asking negative questions: 'In your studies, if you want to identify the needs, ask a positive question, for example: why do you come to our hotels? But if you wish to comprehend their desires, ask a negative question, like why don't you come to our hotels?' This expert believes that, to get the brain's full attention, an e-communication message must be based on four values: 'pleasure, exchange, sharing, performance for the customer and the company'. He emphasizes the importance of being different rather than being the best. To improve the efficiency of e-communication, he recommends referring to disruption or 'blue ocean' strategies (see Chapter 14). He suggests developing messages based on a four-phase concept: 'Be humorous, create a plot, provoke the unexpected, and incite seduction.'

Communication using social media is particularly important for the dissemination of a brand's values. This is an efficient way to make the brand likeable by making it a topic of discussion within the community. The Converse footwear brand has managed to create a plethora of likeable rumours in social networks, which is unfortunately not the case for its competitor Nike, which has sometimes been criticized in these networks for using Asian manufacturers that used child labour.

The idea of communicating on the terrain of internet users via the different social media is easier to talk about than to implement. When it is poorly executed, its application can cause serious harm to the company's image. Using external or internal professionals may prove indispensable. We are witnessing the emergence of a new profession, that of 'community managers' or social media officers (SMOs), bona fide experts in social media, who perform many tasks:

- They select the social media favoured by the customers targeted by the company. While Facebook, Twitter, LinkedIn and YouTube may seem unavoidable when dealing with a large customer base made up of individuals or professionals, other, less universal media can be more suitable for limited targets. Local habits can result in a more specific selection. In Germany, the Xing social network is very popular for certain categories of internet users, as is Copains d'Avant in France.

- They identify the communities of potential interest to the company.

- They optimize the selection of social media favoured by the selected communities.

- They determine how to communicate with the communities, taking into account their habits and the specific nature of the social media they visit.

Subsequently, web professionals strive to encircle the consumer, using the different resources available from e-marketing:

- Thanks to its interactive capabilities, the blog is the pivotal point of the e-communication process. The website is more passive and remains within the catalogue of offers. They must both be designed based on the customer's issues, not the company's. Customers must recognize their expectations after a few clicks so they can then be pulled in by the company's proposed products and services.

- A good viral video is a useful plus, as it helps establish a sympathetic contact by making customers laugh because of its madcap humour. It must be sufficiently funny and appealing to be worthy of a recommendation to other community members.

- Appeals to visit the website or blog to find relevant information are made via targeted e-mails, avoiding any form of intrusion.

The internet encirclement policy will be consistent with advertising and event-related communication.

All communication and e-communication policies must help establish a permanent interactivity between the company's brand and its customers. With this familiar approach, by taking part in conversations in social networks, the brand follows the objective of being progressively perceived as a new friend by community members.

Recent research in neuroscience indicates a switch from the notion of individual brain to that of collective brain, striving to synchronize the brains in sensory harmony. Music is probably the most efficient tool for creating actual communities of listeners, using the common perceptions and feelings of different brains for certain sounds or musical themes. Other senses can also be solicited to achieve a collective synchronization of the brains and create genuine sensory communities.

At a time when marketing is so keen on community approaches, directly or via social networks, neuroscience research into brain synchronization at community level could become an interesting topic of investigation for Neuromarketing.

Brand policy to reassure the customer's brain

17

The brand is fundamental for reassuring the customer's brain. More than just a name and logo, it represents a number of values associated with a company. Interest in these values leads the customer's intelligence to favour one brand over another, independently of the traditional brain reflex, which is to favour proximity or the cheapest price.

In addition to appealing to customers, the brand is an asset for the creation of a culture based on shared values and a unifying corporate philosophy. It is becoming a key element in mobilizing all the personnel of head offices and networks at national and international level. When these values are respected, by sharing the company's social policy with the staff, it constitutes a genuine anti-stress tool.

By making promises to target customers, the brand is a decisive weapon that helps reassure and retain consumers. By creating strategic coherence and unifying personnel, it creates value and constitutes veritable capital for the company.

The Interbrand company publishes the ranking of the 'most powerful' brands every year in the *Financial Times*, as well as their monetary value. The clear winner is Coca-Cola, whose brand value alone is estimated, depending on the year, at approximately €70 billion. It is usually followed by Microsoft, estimated at €65 billion, and IBM at €55 billion. European brands include Mercedes (€22 billion), LVMH (€8 billion) and L'Oréal (€6 billion). These estimates are provided on an indicative basis. They vary from year to year, notably in times of crisis.

According to Jean-Noël Kapferer (1994, 2008), a world expert on brands, 'the brand is a powerful name': a power to influence the customer, but also the company's partners and the environment. It is all the more necessary as the customer's brain is exposed to significant risk in a purchase situation. It reassures the brain and promotes loyalty.

There are multiple issues at stake for the brands:

- Give the brain of existing and potential customers objective reasons to choose a retailer independently of price and proximity criteria.

- Convince the customer's intelligence to subscribe to the products and services of a company or buy from its distribution channels by embracing different values to those of the competition.

- Confront new entrants. The presence of a brand with strong professional attributes is a major barrier that helps prevent 'outsiders' from penetrating the market. A well-known brand reassures the customer's intelligence. When it is not well known, the brain attempts to protect itself against the risk of the unknown and dissuades the customer from buying.

- Withstand the pressure of overly demanding partners, notably when they distribute the products and services. Companies involved in the consumer goods domain are well aware of the importance of the brands when dealing with the many demands of hypermarkets' purchasing units.

- Create shared values for all the personnel of a group, in particular in the context of restructuring and international development operations. The context of merging businesses raises the question of whether the management wishes to create a corporate culture common to all subsidiaries or to let each subsidiary live its own life as an independent company. The desire to create values common to all group subsidiaries was a major factor in the choice of a unique global brand for AXA, Total, Allianz, Suez and Vivendi.

- Make a fresh start subsequent to recent events the company wishes to forget. Worldcom became MCI; Vivendi Environment turned into Veolia to show its break with the Jean-Marie Messier era; Crédit Lyonnais changed its name to LCI.

Creating a brand to reassure the customer's brain is far from easy. Brand development must be guided by a policy, followed by resources and managed by an organization. Ultimately, the most important thing is not what the brand says to the consumers but what consumers say about the brand. Based on his survey of 2,081 patients studied with MRI, Lindstrom mentions in his book *Buyology* (2010) that famous brands light up the same area of the brain in their followers as that lit up in a sample of nuns when asked about God, Jesus Christ or the Virgin Mary.

Define a policy to complete the triad of positioning–identity–brand

The triad: a progressive approach to brand policy

Brand policies are developed in the form of a progressive process. As shown in Figure 17.1, it involves the choice of a positioning, followed by an identity and finally the brand.

FIGURE 17.1   Positioning–identity–brand triad

Positioning represents the strategic desire. It corresponds with the values the company wants to convey to the customers' intelligence via a future corporate brand, or with the promises attributed to a brand chosen for a product, service or sales network.

The identity concerns the personnel of head offices and distribution channels. It consists of acquiring corporate values and creating osmosis between the personnel's behaviour and the values advocated by the positioning. The Nike company, for example, strives to create symbiosis between sporting values and the behaviour of all its personnel.

The brand becomes a reality when the company's values and promises are perceived as real and distinct from the competition by the intelligence of existing and potential customers or by that of other players such as shareholders, future recruits, staff from distribution networks, and prescribers.

Brand positioning

Positioning corresponds with a strategic choice for the company. According to Kapferer (1998), it must answer four key questions: Why? For whom? Against whom? For when? We will illustrate this with the example of the policy used by the AXA insurance group to create a unique international brand:

- *Why?* To define one's own identity based on specific values, to stand out from the main competitors, to orient the entire group based on shared values, and to create, in each national and international subsidiary, a feeling of belonging to the company and sharing AXA's values, as stated by Olivier Mariée, AXA's Marketing and Distribution Director during a conference at HEC Paris.

- *For whom?* For existing and potential customers, the personnel of head offices and networks, shareholders, financial analysts, future staff, etc.

- *Against whom?* Rival insurance companies, but also new entrants, and banking networks, the retail industry, and websites selling insurance and financial products and services.

- *For when?* It varies depending on the country, on whether or not the local brand is strong. Time required to switch to the unique brand: five years.

Brand identity

How does one differentiate between BNP-Paribas, Société Générale and Crédit Lyonnais? How does one distinguish between Banque Populaire and Crédit Mutuel? For insurance companies, what are the unique characteristics of AXA, GAN or AGF, Groupama or MMA? Obviously, the notion of positioning has its limitations: while it helps differentiate between companies targeting different customer groups, with different products or services, thereby enabling basic differentiation, it is of no help to communication managers aiming for the same targets, with the same resources and same services. This situation is typical of the competition in many areas of activity. To address the limitations of the positioning concept, Kapferer (1991, 1998) and Variot (2001) propose an original operational methodology to identify where the unique features of a company lie. Their approach is based on the concept of brand or corporate identity: they call it 'the identity prism', as presented in Figure 17.2. It directly involves the different representations of the customer's brain when the brand of a company, product, service or distribution channel is mentioned. These representations are linked to what Antonio Damasio (2005) calls the 'somatic markers of the brain'. These markers are linked to the brain's recording of all the major happy or unhappy experiences every customer has had in the past. They

FIGURE 17.2 The identity prism

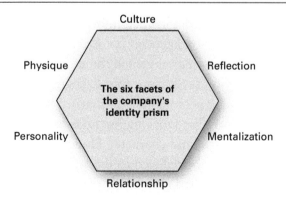

SOURCE: Inspired by Kapferer (1991, 1998) and Variot (2001).

can appear subconsciously when the brain is confronted with the mention of a brand. The positive or negative effect of these memories linked to the mention of the brand can result in subconscious reflexes of acceptance or rejection. The brand's identity, through its codes, history and sensory evocations, must be designed to be perceived positively by the brains of consumer groups of interest to the brand.

According to Kapferer and Variot, the identity is made up of six facets: *physique, personality, culture, relationship, reflection* and internal *mentalization*. These facets form the identity prism of every emitter, whether it is a company or a brand. The difference and durability of any retailer can be encapsulated in one or several of these six facets. We will tackle each of these identity facets:

1 *The physical facet* represents the characteristics of the company, its resources, products or services. This is the traditional realm of communication, of 'product strengths'. It is also a facet that can be easily standardized, as services are very similar (Variot, 1982).

2 *The 'personality' or 'character' facet* corresponds with an anthropomorphic vision of the company, a personification of the emitter. If L'Oréal, Ikea and Toys R Us were people, how would they be described by human intelligence in terms of gender, age, style, social status, character traits or personality? Certain companies are perceived by the consumers' brain as 'serious', others as 'smiling' and yet others as 'dynamic'. This identity facet has been included in the working methods of many communication agencies for years. According to the Ted Bates agency, which invented the notion of the unique selling proposition (USP), 'product strengths' also refer to the 'unique selling personality'. Similarly, Séguéla (1982) organized his agency's working methods based on the two notions of physique and character. However, the identity of a company cannot be reduced to these two facets, which only specify the 'constructed emitter', ie the speaker. We will now tackle two other identity facets that concern the 'constructed recipient'.

3 *The reflection facet is based on a mechanism designed to identify human intelligence.* All retailer communications implicitly stipulate what type of individual they are talking to. We are not referring to the target (the objective recipient) but to a constructed recipient, the reflection of the brain's own image (even if it does not correspond with the objective description of the customer). Saying 'a company for successful people' means constructing a specific recipient, providing the intelligence of target customers with a certain image of themselves.

4 *The mentalization facet corresponds with the 'internal' aspect of the constructed recipient.* While reflection corresponds with the external mirror, this facet refers to the internal mirror. Deep down, how does the customer's brain see itself when in contact with the products or services and distribution sites of a retail company? Is the customer's intelligence encouraged to intimately perceive itself as wise, thrifty or avant-garde, as a pioneer, a shrewd manager or a considerate and attentive father?

5 *The relationship facet corresponds with an observation:* all communications automatically propose a certain type of relationship between the recipient's intelligence and the emitter, or a certain type of rapport. The nature of this rapport, implicitly proposed, constitutes a key facet of the identity. It can be a rapport based on education (the retailer teaches you something), partnership (the customer is a partner), complete care, domination, mothering, admiration or connivance. This facet helps put into perspective the notion of 'star strategy' developed by Séguéla: considering the company as a star is proposing a relationship based on admiration or even worship to the recipient's brain. This type of relationship however is not the norm. Its objective is to simplify the brain's responses via the imitation phenomenon, as developed in Part I. Communication, as stated above, can propose many other types of relationship between the company and the recipient. Van Aal (1981) claims that communication constantly creates connivance between the emitter and the recipient's intelligence. This is once again an excessive generalization: connivance is one type of relationship among many. Every company must carefully define the type best suited to itself and its audience.

6 *The cultural facet refers to the cultural roots of any company.* Via the company, the customer's brain accesses its universe and myths. Owning a chequebook from the Rothschild Bank means symbolically appropriating part of the Rothschild saga. Being a Barclays customer also means embracing signs of the British Empire, of the City, etc. Therefore the identity is also a part of the cultural facet: what universe do we access through the company? Manhattan, Europe or France? What myths do we access? In an interesting book, Lewi (2003), another expert on brands, compares brand identity with Greek gods; he claimed, during an HEC Paris conference, that brands that are sustainable over time are those that refer to the great founding myths of humanity. These myths have profoundly influenced our intelligence, mostly subconsciously. When a brand refers to a myth, the brain awakens its knowledge of it and stimulates the interest in the brand by comparing it with the myth buried in its subconscious.

The combination of these six facets defines the emitter's identity. Taking into account only one or two of these facets, as is still often the case (physique and personality), means leaving any consideration for communication form, tone or style to creative inspiration. At a time when companies have increasingly similar messages in substance (physique), form becomes essential to convey the differences in identity. The customer's brain subconsciously deciphers the very form of messages, the tone of the communication and signs emitted to get a sense of 'who's talking', ie the emitter's identity. The definition of all the facets of the identity prism for a retailer precedes the image strategy and provides a charter that will guide all communication decisions (form, style, tone, codes and substance) without exception.

The brand

The brand is constituted when the values associated with the positioning decided upon by the management, which the personnel identify with, become a reality for the final recipients (customers, shareholders, environment, prescribers, etc).

To achieve this, they must often transcend purely technical or professional aspects in order to take on ethical or social attributes. In the retail industry, FNAC strives to acquire an initiating corporate value to help the customer's intelligence make the best possible choice in keeping with his or her aspirations. Leclerc positions itself as a company fighting a permanent political battle against prices, and Danone as a brand that cares about health and not just about the taste of its products.

In the United States, Nike attempts to create genuine osmosis between its brand and the sports world. Its purpose is to aim for identification between the values associated with the company and the feelings of the customers or shareholders. There are four advantages inherent in the creation of a strong brand with recognized and appreciated values. It reasserts the value of employees and increases their personal worth. It creates confidence in the future for shareholders and improves the credibility of the stock. It reassures the customers' brain and reinforces their loyalty. It activates the potential customers interested in the promises and helps win them over.

To ensure that the values or promises presented always meet the customers' expectations, the brands must evolve as new needs emerge, as expectations change, and as the competition undermines the advantages of the original positioning. Far from being set in stone, the brand policy needs to be managed over time and space. Company acquisitions or mergers with new national or international partners force companies to rethink their brand policy at corporate level as well as across their lines of products and services or distribution channels. The brands must constantly adapt to cultural evolutions and the evolution in the trends that interest the customers' intelligence.

Implementation of the brand policy

The brand strategy

Implementing a brand strategy consists of making decisions regarding multiple questions:

- Should a unique institutional brand, either national or international, be created or should existing brands be kept alive?

- How does one proceed not to decrease value when the decision is made to create a unique and common brand, either nationally or internationally?

- Should common or diverse brands be created when product or service lines are designed for different customer segments or when distribution channels are competing against each other?

Answers to these questions are not universal. They largely depend on choices made by executive committees during the conception of their strategic and marketing plans. They can be diverse: creation of an international unique brand, independence of local brands, proposal of a common logo for the different brands of the group while keeping local brands, fusion of two banner brands, and so on. The choice of the brand can also evolve depending on the realization of important strategic projects, or environmentally linked problems. TotalFinaElf was renamed Total after the successful completion of the merger between the three firms.

If choices have to be made by each firm, many experts believe that respect for a few principles may improve the efficacy of the brand strategy. Here is some of the experts' advice:

- A national or international unique brand is of interest when a group strives to acquire shared values, or wishes to present a global and coherent strategy to shareholders and reassure the customers' intelligence, exhibiting an international power. In such a context, a strategy like the one adopted by AXA, which lies in its famous 'Think global, act local' claim, may appear interesting. Conversely, if the executive committee wishes to present the acquired firms as independent, the use of different brands appears more appropriate. This is the case for Société Générale with Crédit du Nord, of Crédit Mutuel with CIC, and of Caisses d'Epargne with the Crédit Foncier de France.

- When a group decides to create a unique brand, implementation should remain pragmatic in order that the decision is perceived as inevitable by all firms of the group. It should be particularly the case when the new brand is substituting for nationally strong brands. Having put the two brand names side by side, the abandoning of the old one for the unique new one should be done only when the notoriety obtained in the customers' brain of each country has reached a sufficient level. In the insurance sector, this progressive strategy has been adopted by AXA to replace well-known local brands such as Colonia in Germany, Royale Belge in Belgium and Sun Life in Great Britain. The deadline for the final change of brand names can't be decided completely in advance for fear of decreasing value for the stakeholders of the concerned countries.

- Managing many brands is very costly for firms. The trends in fast-moving consumer goods are towards an important decrease in their number. Keeping different brands can only be justified when the firm targets different consumer segments or different distribution channels or offers very different products or services. Selling similar products or services at different prices by the same firm across different channels, which may be in competition, can justify the creation of distinct brands. This is frequently the case in the insurance sector, which commercializes the same products across different channels: general agents, home sellers, brokers and direct insurance.

The marketing mix of the brand

The marketing mix of the brand focuses on implementing and harmonizing the different actions that will convey communication or the values and promises advocated by the brand.

FIGURE 17.3   The marketing mix of the brand

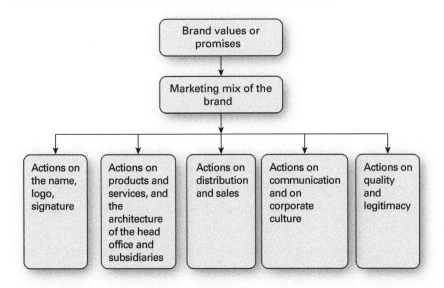

Name, logo, signature

The customers' intelligence associates the brand with the elements that characterize it: a name, a logo, a signature, etc.

The first identification and distribution element is the name. It must capture the imagination of everybody's brain while conforming to the identity that the company wants to give it. The idea is to find a name for the company but also for the products and services proposed, or for the different distribution channels used. When the company is involved on the international market, the name must be acceptable in all countries and easy to pronounce in all languages, which is no small task. The search for a name can be extremely complicated, as trends fluctuate, as do customers' reactions. Furthermore, once a name has been found, after overcoming many difficulties, you must make sure it is available, which is far from easy. The services of specialist agencies are indispensable for finding brand names corresponding with the guidelines determined and to find a logo and a signature.

The brand name is generally accompanied by a logo, which will adorn business cards and all other documents used to communicate.

This logo must also express the company's values as faithfully, concisely and creatively as possible. The graphic charter presenting the colours, shapes and symbols chosen reflects the particular meaning that the organization intends to convey to the customers' brain.

We will examine the example of the BNP-Paribas logo. As pointed out by the Communication and Advertising Manager (Rancev, 2002):

> The idea behind this logo is to promote the new strategic impetus born of the merger with Paribas. We can see the BNP-Paribas stars turning into birds along a dynamic trajectory. The star symbolizes the guide and is a reminder of the fact that BNP-Paribas belongs to the European Union. The bird is ready to travel the world and to land, as the account manager is ready to assist customers in their future projects. Green is the colour of hope and gives the ascending curve notions of transparency and the environment. BNP-Paribas deliberately stands out to strongly assert its desire to innovate, its ability to anticipate change and its ambition to open up to the world.

The signature highlights the message the retailer wishes to communicate to the intelligence of its existing and potential customers in connection with its image. Signatures evolve over time. They have moved from a self-centred perspective in the late 1980s to getting closer to consumers, making them the focus of the relationship. The approach consists of changing 'we' to 'you' and ultimately 'I' to involve the customer as much as possible.

The signature is a major feature of the identity and brand. To increase its efficacy, this signature can involve the primary senses of the customer: sight with written elements, like Obama's 'Yes, we can' or Nike's 'Just do it'; hearing based on the musical jingle, like the song for the MAAF company; and smell, by associating it with the logo, which is increasingly common.

Uphold the reputation of the brand by adapting the marketing mix

To uphold the reputation of the brand, adapting the actions of the marketing mix is a necessity. This means designing product and service offers that will keep the promises made. This also concerns the architecture of the head office and sales outlets. Darty invests significant amounts in its after-sales department to keep the promises of its positioning. The appearance of Hermès, Chanel or Vuitton stores reflects a high level of luxury.

Personnel from physical distribution channels may also need training and must change their behaviour to achieve the ambitious objectives determined by the positioning. The online content cannot disappoint the intelligence of internet users, in particular when the brand aspires to modernity. Social networks can make the brand pay dearly for any discrepancy observed between its perception and its actions. Nike, accused of being an exploiter, is criticized for its Chinese factories. By contract, Converse enjoys a good reputation among teenagers, who publish photos of their decorated trainers on their blogs.

Internal and external communication is a key success factor in getting the intelligence of customers, shareholders and various inter-mediaries to adhere to the attributes of the brand. It is necessary for launching actions designed to create a corporate culture in keeping with the ideas championed by the brand in order to create osmosis between the perception of the customers' brains and the values advocated by the employees of the company.

The positioning of the different retailers in terms of ethics or sustainable development can only be credible if the personnel behave in such a way as to keep the promise strongly stated by communica-tion. The Nike brand encountered huge difficulties when the com-pany was accused of supporting child labour in Asia.

A series of in-depth actions relating to quality and legitimacy, in all domains of the marketing mix, is a necessary corollary to ensure the success of the positioning–identity–brand triad in the customers' brain.

Brand sensoriality

As part of his extensive research conducted using the MRI in the United States, Lindstrom (2010) explores how the brains of specific categories of individuals react to certain dialogues or information received. After studying a population of nuns, he noticed that a very specific part of their brain, an area that is normally dedicated to joy, serenity or love, lit up when he mentioned names like God, Jesus and the Virgin Mary. When he continued his research, he was surprised to find that the same area of the brain lit up in brand aficionados or 'addicts' when he mentioned this brand or presented some of its characteristic attributes. A strong brand is a brand that does more than just respond to consumer needs or expectations: it creates emotion in the brain. Lindstrom (2010) states: 'A brand is pure emotion; it cannot be seen, smelled or touched.' This emotion can turn into affection, bordering in some cases on the idea of elevated distinction, and can sometimes take on an almost mystical sense.

Some people are prepared to spend tens of thousands of euros at auctions to buy a traditional leather bag, as long as it is a Hermès Kelly bag. Certain 'addicts' do not hesitate to spend hours waiting, sometimes all night, to buy the latest Harry Potter novel or to be the first to purchase the latest Apple device. By largely impregnating the brain, a powerful brand has an influential effect on the customers' decision to buy.

Creating an attractive brand is far from easy. An attractive brand must convey the impression that it is not limited to doing something but that it is trying to bring something to life. It must create enough affection so that consumers buy its products, not only because they like them but also because they like this brand and they wish to belong to a community of members devoted to this brand. By directly accessing the brain, the senses, when properly solicited, contribute to creating a familiar and affectionate experience with the brand. The success of a brand policy consists of creating but also sustaining what Maurice Levy, CEO of Publicis, refers to as a 'familiar relationship with the brand'.

Somatic markers at the service of the brand

The somatic markers discovered by Damasio (1994) correspond with memories relating to acquired preconceived notions or significant events deeply rooted in the brain. They are frequently linked to one of our five senses and can re-emerge subconsciously when this sense is solicited.

The European consumer's brain often associates a notion of good technical expertise and reliability with a German-sounding brand without even thinking about it.

By choosing the name Weston, the famous French footwear brand, founded in 1891, intends to evoke British chic. By using the name K-Way, the windbreakers of the Duhamel company, also of French origin, give their clothes a notion of American efficacy that is essential in coping with the storms of the West Coast of the United States.

When he smells the scent of a madeleine cake, Proust's (2003) hero recollects the happy moments of his childhood. The scent of the brand can place customers in an ambiance of well-being by reminding them of happy experiences and predisposing them to buy.

Certain colours are also assigned specific meanings. Achromatic colours such as white, black and grey are frequently associated with a luxury or 'top-of-the-range' image. They can reflect different associations depending on the culture or country. White is associated with purity and happiness in Western countries, while it is linked to death in many Asian countries. The colour orange is considered festive in Holland and appreciated in Ukraine, while it has a negative connotation in some parts of Ireland.

Brands can try to connect with somatic markers so that the consumers' brain assigns an emotional image to them. They can also attempt to create somatic markers. Certain markers based on fear, for example becoming obese, getting old, having wrinkles, losing hair and eating junk food, are likely to create a strong attachment in the brain to brands that claim their willingness to combat these issues.

To develop loyalty and become essential, brands must offer personal experiences. They can take advantage of storytelling by recollecting defining moments associated with the emotions of reading or seeing things in the unconscious of the consumer's brain. They can also extend the experience established with customers by presenting them with a story to be told. The brand's story, told in the form of actual or fictional sagas, invites customers to be part of the tale and share the feelings and values of the heroes. Buying Chanel products means subconsciously participating in Coco Chanel's saga. Wearing a Ralph Lauren shirt means being part of the British Empire. Lacoste takes customers into the world of tennis. In Shanghai, Dunhill is creating a sales outlet in an old colonial house decorated like Alfred Dunhill's old home. His story is told to create an object of affection.

Major brands often use, deliberately or otherwise, ingredients that solicit the brain's somatic markers linked to elements with a strong impact on the collective unconscious, eg a secret, like that of Coca-Cola's formula or Nike's air bubble, a bible, like that of McDonald's, which is particularly bulky, bigger-than-life characters like Kentucky Fried Chicken's Colonel Sanders, British and US aviators from the Second World War wearing Mac Douglas or Chevignon bomber jackets.

Sensory marketing of the brand

The purpose of sensory marketing is to use the senses, which have a direct influence on the brain, so as to multiply emotional connections with the brand. Emotional tagging strives to give the brand an emotional personality. In addition to sight, all our senses affect the emotions and can be used to create a more emotive image. This is why the visual logo is often completed by an audio, olfactory or tactile logo.

Sounds can create or transfer emotions and stimulate memorization. In the insurance domain in Europe, listening to the famous French song *La Ouate* means transferring part of its emotional affection to the MAAF brand and improving its memorization. Many companies decide to create an audio logo, eg SFR, Starbucks, Nespresso and Yves Rocher.

Odours also affect emotions. Sweet smells give a sensation of pleasure, while bitter smells tend to irritate or disgust. The neurons of the olfactory area directly transmit information to the limbic area of the brain involved in emotions and affective memory. The olfactory memory is better at withstanding the test of time, which facilitates brand memorization. This is why some retailers are keen to acquire an olfactory signature.

Congruence of the senses of the brand

The brand's sensory policy transforms the customer's experience into emotional and affective addiction, all the more so as it creates harmony between the different senses solicited, which experts refer to as 'the congruence of the senses'. It must subsequently ensure the coherence of all sensory logos through all means of contact and communication with consumers. In addition, all the senses solicited must correspond with the positioning chosen for the brand.

To convey the image of a calm, relaxing, thoughtful company, it is preferable to use pastel colours (pink, blue, green, etc) combined with soft, slow music and the diffusion of relaxing scents (lavender, rose, orange, sandalwood, etc).

An increasing number of brands mention the desire to create a sensory experience for their customers:

- Luxury cosmetics brand Decléor, which belongs to the Shiseido group, specializing in aromatherapy based on essential oils, is particularly eager to develop this type of sensory experience with the consumers of its products.

- Nespresso wishes to give its brand an image of rarity, exclusivity and luxury. To do this, it has opted for top-of-the-range stores, select addresses reserved for club members, a limited edition of its luxury magazine, and a communication strategy that reinforces the notion of selectivity.

- The famous New York chocolate maker Hershey's, in an effort to allow chocolate lovers to enjoy a genuine sensory experience with the chocolate and the brand, opened a store worthy of Tim Burton's famous film *Charlie and the Chocolate Factory*. Hershey's Chocolate World enables visitors to enjoy a complete chocolate experience, going as far as creating their own chocolate bar.

- For their clothing brand, Abercrombie & Fitch establishes a specific atmosphere based on a combination of sensory stimulants, including lighting, music, olfactory signature, and social representation based on the sartorial appearance of the sales staff.

Retail brands, as developed in our chapter on sensory marketing in the sales outlet, are increasingly keen on these new approaches emanating from sensory marketing and Neuromarketing. The approaches are not limited to luxury brands. In Belgium, the 'leader' of discount outlets, Colruyt, wishes to provide its customers with a sensory experience based on values relating to asceticism, frugality and smart shopping. The brand essentially targets consumers who have values other than consumption and who are not susceptible to excessive solicitation. To convey this impression, the sales areas are designed with undecorated flooring, neon lights, products presented in transport boxes, and refrigerated units replaced by cold rooms.

Companies' growing interest in creating emotional brands, in an increasing number of areas of activity, means that sensory marketing and Neuromarketing are destined to play a key role in tomorrow's marketing.

18 Quality to enhance loyalty, and legitimacy to leave the customer's brain with a clear conscience

The purpose of Neuromarketing, as described by the Neuro-marketing method, is not simply to sell to the customer's brain but also to turn the customer into a partner. It wants the brain to allow the company to communicate regularly with it as part of a 'permission marketing' approach. In addition to winning over customers, Neuromarketing focuses on satisfying the customer's brain within the context of this partnership. This is a prerequisite for retaining customers and increasing their value. The quality policy constitutes a major asset in this approach. Its marketing definition is simple: 'Quality is when the customer returns, not the product.'

Apart from quality, 21st-century consumers expect even more from their partner suppliers. Their brain wishes to have a clear conscience whenever they buy. Customers are increasingly keen to consume healthy products, taking an interest in organic agriculture, non-polluting products manufactured based on sustainable development imperatives and fair trade products.

They do not wish to become the partners of companies that do not treat their employees and staff respectfully. Following the announcement concerning the many suicides within France Telecom, its mobile telephony subsidiary Orange apparently received several cancellation requests from customers unhappy with the parent company's behaviour.

Beyond quality, consumers want to establish regular relationships with companies that display a sense of ethics, are socially responsible, and meet sustainable development requirements. Beyond the quality offer, Neuromarketing must help companies acquire genuine legitimacy in the eyes of their customers. This legitimacy must respond to ethical behaviour standards inside and outside the company.

Being socially responsible vis-à-vis the personnel, the environment or the country is becoming a quality that appeals to the intelligence of proactive consumers. Traditional ISO standards attesting to a high level of quality will soon be completed with ERGO standards, certifying that the level of stress within the company is acceptable for the staff.

With social networking, proactive consumers are becoming information producers. Any problem in terms of behaviour or quality is rapidly exposed on blogs and can spread in a matter of days to thousands of people who are part of these networks.

Quality and legitimacy: an imperative for Neuromarketing

Quality to satisfy the brain and retain the customer

Once the sale has been completed, a quality policy is necessary to satisfy the customers' brain by reassuring them about their purchases. In addition to winning over customers and selling, marketing also contributes to enhancing customer loyalty. This is essential for at least two reasons. The first comes from the fact that products and services cannot be intensely promoted among customers without an appropriate loyalty strategy. This is the purpose of the policy intended to intensely cultivate customers. The second reason relates to the idea that a fully satisfied customer often constitutes the best possible communication vector for the organization. According to Collet, Lansier and Ollivier (1991), you often hear that 'a happy customer generally shares his satisfaction with three to four friends, whereas when he is unhappy he often tells more than twelve people about it'. The saturation of numerous markets and the limited renewal of customers as a result of the effect of demographics naturally emphasize the importance of this issue. The solution can only be found in the implementation of a wide-scale policy aiming for total quality confirmed by suitable ISO certifications.

FIGURE 18.1   Quality to enter the virtuous circle of loyal customers

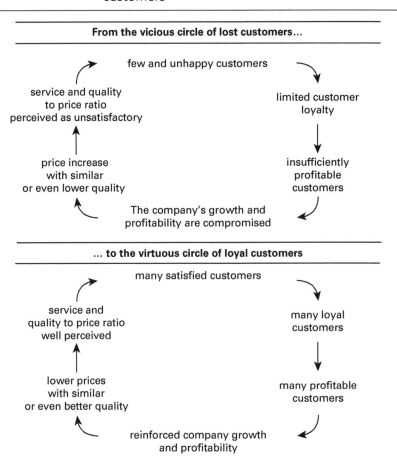

From the vicious circle of lost customers...

few and unhappy customers

service and quality to price ratio perceived as unsatisfactory

limited customer loyalty

price increase with similar or even lower quality

insufficiently profitable customers

The company's growth and profitability are compromised

... to the virtuous circle of loyal customers

many satisfied customers

service and quality to price ratio well perceived

many loyal customers

lower prices with similar or even better quality

many profitable customers

reinforced company growth and profitability

SOURCE: Inspired by Moiroud (1993).

The implementation of a quality policy, as shown by Moiroud (1993), can help the company leave the vicious circle of lost customers and enter the virtuous circle of loyal customers.

In his book *The Loyalty Effect*, which has become a worldwide best-seller, Reichheld (1996) stresses the need to permanently satisfy customers. He writes: 'The best way to sustain growth is to look after happy customers and always ask them the decisive question: would you recommend our company, our products to your friends?'

Yesterday's reticence becomes tomorrow's obligation

The search for quality has already proved its worth in other latitudes and other trades. Quality circles, a tangible emanation from this policy, were born in Japan in the 1960s, crossed the Pacific en route to the United States around the 1970s and reached the European shores *circa* 1980. From then on, their development was exponential.

Companies' attitude to quality is bound to become a concrete reality for at least three major reasons: *profitability*, *competitive pressure* and *the evolving expectations of the customers' intelligence*, especially when participating in social networks.

Firstly, with regard to profitability, Raveleau (1990) points out that, 'if companies kept a record of all unnecessary, poorly executed and repeated activities, this would equate to 20 per cent to 30 per cent of the turnover'. Other experts estimate the cost of non-quality at a more modest 10 or 20 per cent depending on the company. Merrill Lynch has set up nearly 200 quality teams to reduce errors and simplify procedures. According to the company's managers, this has generated cost reductions of up to €5 million. These savings are undoubtedly a significant source of reflection for companies strongly aware of the need to achieve productivity gains.

Competition is probably the second reason for the companies' interest in quality. In many industries, such as automotive, telephony and services, the lack of quality in relation to better competitors will probably become a major cause of decline.

The quality asset is all the more crucial because customer expectations are also moving in this direction, as witnessed on a daily basis by networks in regular contact with customers. More 'multi-equipped' and increasingly solicited customers are no longer inclined to accept the constraints that, until recently, were imposed as unavoidable. A poor greeting, a long acceptance period, incompetence in terms of advice or information, computer errors, reluctance to reimburse, bad faith and saturated call centres are becoming pretexts for terminating contracts and dropping suppliers in a context of fierce competition. Customers are becoming more and more sensitive to incidents. The development of the internet in the form of forums, chats and various ways for customers to meet and expose the difficulties encountered in their dealings with institutions only reinforces the need for quality.

Total quality: pastime or ultimate goal for Neuromarketing?

As pointed out by a quality manager in an international chemicals group, 'Focus, dialogue, customer respect, internal and external consultation to find the right responses to actual problems are quality factors which require a genuine cultural approach within European businesses.' This approach, difficult to implement, must initially resist the temptation to take the easy way out and must be wary of gadgetry. The quality mindset will permeate through the company in a profound and sustainable manner only if it is based on an actual strategy, supported by appropriate resources. Initiatives following the trend of appointing a quality person or developing quality circles, interesting though they may be, are not enough. They may ultimately be ineffective if they fail to fit into the context of a global approach.

The concept of total quality can be considered the ultimate goal as part of a successful Neuromarketing policy. Ceaselessly promoted by the management of the organizations, it requires the profound modification of the mindset as well as the full commitment of all personnel to customer satisfaction. The concept of total quality must become a source of pride for companies. It relies on the famous principle: zero defects, zero delay, zero stock, zero paper wastage and zero incidents, along with the addition by some experts of zero accidents and zero contempt.

ISO certification, which customers are increasingly requesting, and commitments to consumers in the form of charters are undeniable assets that help retailers enhance the level of quality and render it sustainable. This sustainability must be maintained over time but must also be standardized throughout networks and subsidiaries.

From quality to legitimacy

The change in mentality linked to the growing influence of the environment is forcing companies to develop policies exceeding the scope of quality and aiming for legitimacy. Having a clear conscience and being legally correct are no longer sufficient for retaining customers and winning over potential customers if the retailers' attitude is not considered legitimate by the environment. Beyond the quality of the offers, commercial relationships and communication, the consumer's intelligence is more and more sensitive to the ethical, social and societal role of the businesses it wishes to deal with. Legitimacy within the environment concerns all proposals and actions, extending beyond the mere customer–supplier relationship. A bank financing polluting industries, the refusal to insure certain accidents that cause misery to a population as a result of unforeseen disasters, the anti-social behaviour of managers or personnel, and the form of communication can all be considered illegitimate by customers and cause them to reject the retailer's solicitations. These phenomena can have an even more negative impact if part of the image and identity of the institution is based on an ethical approach. The Benetton company experienced a vehement rejection of its communication methods in the United States, deemed shocking and illegitimate by its customers. Faced with the loss of customers and distribution networks, it was forced to thoroughly review its communication policy. This policy involves the behaviour and attitudes of all managers and employees when they are assimilated with the company's image.

As well as quality, Neuromarketing must contribute to consolidating the retailers' legitimacy. It must pave the way for the notion of a sustainable and socially responsible company.

Organization of the quality–legitimacy policy

Total quality: a new mindset

To be viable and effective, a quality policy must achieve three objectives: *improve customer satisfaction*, *increase productivity* and *reduce costs*. This desire requires a profound change in the companies' mindset.

The first change aims at making the primary concern of your workforce that of satisfying the customers by resolving their problems. This is one of the key concerns of Neuromarketing. The second consists of acknowledging that quality, like legitimacy, concerns all departments and all employees from the top down. The malfunctioning of a single cog can affect the entire policy. This change requires the implementation of a trust culture and a genuine participatory management system. As pointed out by the managing director of a chemical company, 'The new organization must rely on the firm conviction that freedom works better than constraint.' This requires the progressive reduction of Taylorism and bureaucracy within the company, and the education of staff in quality and ethics. With this in mind, one Swiss insurance company, eager to develop a total quality policy, is experimenting with a three-phase process:

- Firstly, improve the employees' friendliness across the different departments of the company by creating an atmosphere of communication, mutual understanding and trust within the institution.

- Secondly, develop the most courteous and competent relationships between the personnel of the head offices, staff of the regional branches and network employees (general agents, brokers, in-house producers, etc).

- Thirdly, implement a proactive policy promoting competence, courtesy and dynamism between company representatives (head office and networks) and customers. This policy as a whole, combined with the improvement in technical and commercial procedures, must ultimately allow the Swiss insurance company to tangibly legitimize its positioning as a 'courteous, professional and dynamic' company.

Quality or legitimacy policy for a different view of the company

The development of a new mindset, a prerequisite for any quality or legitimacy policy, is difficult to achieve if it is not supported by the appropriate organization of the company. Amateurism in this domain can have a more detrimental effect than the status quo.

This organization is primarily based on the appointment of a quality or legitimacy manager. Reporting to the management, this manager must have strong credibility and an excellent reputation among central directors as well as the personnel. While many companies have entrusted the marketing director with this task, this is not an obligation.

The second objective generally consists of creating two quality or legitimacy committees, an advisory committee and an action committee. The purpose of the advisory committee is to examine the development of the policy per se, define the guidelines of the long-term strategy and propose the necessary resources. It is responsible for submitting a quality improvement plan to the management. It can also be involved in the choice of ethical or societal positioning with a view to consolidating the retailer's legitimacy within its environment. The function of the action committee is to implement the quality policy in accordance with the options selected by the management, as well as to supervise its application in the different departments concerned.

The third objective of the organization consists of appointing quality–legitimacy coordinators within the different departments. Depending on the size of the problem, this can be the quality manager's superior or a manager selected to represent the superior. Whatever the case, it is imperative that these employees are directly responsible for the evolution of quality or legitimacy in the department that they represent. They will eventually engage in dialogue with the representative trade unions to get their opinion or approval of the legitimacy policy aimed at creating a socially responsible company.

The final objective is to implement a number of elements with a view to mobilizing the personnel, while monitoring the quality or legitimacy evolution process by controlling the results achieved. Once again, the creation of adapted structures is a necessity: quality or progress circles, in accordance with the well-known method of Japan's Ishikawa, training and mobilization of directors and personnel, internal information-sharing systems, realization of studies and surveys to monitor the evolution of results, surveillance of conversations on blogs, and community and social networks concerning consumers' perception of quality.

The success of a quality and legitimacy policy is always a long-term process that is difficult to achieve. Two academics from Quebec, Chebat and Langlois (1989), propose a method designed to implement an effective policy in this domain, based on what they refer to as the five Cs: *communicate, commit, cooperate, create a quality environment, create a quality culture*. Communication is the cornerstone of the success of Ford's famous campaign among its employees with the slogan 'Quality is job number one'. It is also considered a key success factor in the policy undertaken by First Security, which as part of a positive campaign reasserted the value of its personnel by showing shrewd sales representatives resolving difficult problems posed by their customers. Committing consists of entrusting a prominent leader with the quality policy.

Quality and legitimacy to enhance loyalty and sponsorship

Quality policy and legitimacy strategy are becoming key elements in retaining customers. This is the purpose of customer relationship management (CRM), relational marketing based on increasingly complete customer information, and the one-to-one approach discussed in Chapter 16. Beyond the quality of products and services and the commercial relationship, a growing number of customers feel more attached to a company if they share its fundamental attitude to the environment and society. In addition to contributing to the country's economy by paying taxes and employing personnel, certain customers demand that their partner companies be good corporate citizens and socially responsible. They appreciate their initiatives such as financial contribution and active involvement in social, humanitarian, cultural or ecological causes within their local environment or in a broader context. Quality and legitimacy through the image of corporate social responsibility constitute solid foundations on which a loyalty policy can be built. Furthermore, they can lead to sponsorship by satisfied customers or recommendations by community leaders via viral marketing. These two policies are undeniable assets that help support a positive image of the brands in the brains of proactive consumers.

Sustainable development to leave the customer's brain with a clear conscience

The notion of sustainable development

A company's legitimacy can be reflected in its adherence to the principles of sustainable development and corporate social responsibility.

In the introduction of her thesis, Balmadier (2003) shows that companies are slowly but steadily moving towards the notions of corporate social responsibility and sustainable development.

For two centuries, the Industrial Revolution has helped provide Westerners with more comfort (eg heating, hot water, drinking water, healthcare), more products (eg food, clothes, hygiene), more leisure (eg games and sports) and more opportunities (eg car and plane journeys, communication, information). Industrial development has also enabled individuals to earn a salary so that they can benefit from all these productions.

Since its modest beginnings, the company has widened its circle of influence:

- The paternalistic company looked after the life and well-being of its workers and employees.

- As a 'citizen', it became part of the local social fabric, providing jobs and paying local taxes, while participating in the region's development.

- Nowadays (or rather tomorrow), the socially responsible company must take into account all the effects of its activity on the environment in the broad sense: nature, people and society. It must take measures to respect and enforce the principles of sustainable development.

At the end of 1987, the Norwegian politician Gro Brundtland gave a definition of sustainable development in the eponymous report submitted by the World Commission on Environment and Development to the United Nations General Assembly: 'Development which satisfies the needs of current generations without compromising the ability of future generations to satisfy their own needs.'

Only in 1992, during the Earth Summit in Rio, did sustainable development become more than a concept. Governments began to reflect upon concrete objectives in the different environmental, social and economic domains, and 178 of them signed a commitment to integrate measures promoting sustainable development into their legislation. The concept became an imperative not only for governments but for companies as well.

In 1997, in Kyoto, a major concern came to light: climate change, primarily due to greenhouse gases (GHG). A protocol was signed on this occasion, designed to force governments to implement measures to reduce greenhouse gas emissions. A European directive was drafted to impose initiatives upon companies.

In August 2002, a World Summit on Sustainable Development (WSSD) was held in Johannesburg, attended by a large number of companies (eg Areva, Suez, Renault, ABB, DuPont, BMW) and non-governmental organizations (NGOs). Companies began to realize that this reflection was not just hot air from the public authorities, and that being passively subjected to the legislation was not a viable solution in the long term. NGOs were now conscious of the expertise acquired throughout their 'battles' and of the renewed attention paid to their increasingly concrete and credible propositions. Trade unions such as CISI and TUAC, two worldwide organizations, also attended, representing another major element of pressure, as they lived within the company.

The Rio and Johannesburg summits considerably raised the profile of sustainable development (SD). Sustainable development is getting a more and more positive response from the citizens of a growing number of countries, who are being made aware by pioneers like Al Gore and Yann Arthus-Bertrand of the need to preserve our planet. Many are quoting the famous line from Saint-Exupéry (2000): 'We do not inherit the Earth from our ancestors, we borrow it from our children.' To allow their brain to have a clear conscience, they want their purchases to be consistent with this new mindset. They want their partner companies to join them in this process. Companies' commitment to sustainable development is far from seamless. Neuro-marketing must play a key role to help them develop a proactive policy in this domain, initially among their employees by limiting stress, and then in relation to their environment. This condition will be more and more indispensable if the company is to acquire genuine legitimacy in the eyes of customers.

The effects of sustainable development on the marketing of tomorrow

The marketing consequences of sustainable development result from the fact that customers, whether they are individuals, professionals, institutions or companies, support the causes advocated. Relayed by powerful prescribers who are becoming aware of their influence and are evolving into communication professionals (NGOs, unions, etc), sustainable development issues, by allowing the brain to have a clear conscience, tend to guide the consumers' choices. They are found in a growing number of domains that directly concern marketing, as detailed below.

The choice of products and services

In addition to quality, customers also want their products and services to be 'ethical' (ecological to respect the environment, recyclable to limit waste, organic to be in harmony with nature, fair trade so as not to discriminate against poor countries, and in keeping with the principles of acceptable social behaviour to respect the individual).

These demands on the end product are felt throughout the production and distribution chain. It is no longer enough for the company to sell good products. These products must be manufactured in accordance with the principles of sustainable development. The company must control its entire chain of suppliers and service providers as well as the quality of the means of production and distribution used, in a broad sense: raw materials, respect for the environment, and social and financial ethics. The brain of new proactive consumers is becoming more interested in companies' responsible and ethical attitudes every day.

The choice of investments

If the customers' reaction is numerically significant, it contributes to the success or failure of a company. On the volatile financial markets, a company losing its customers runs the risk of being downgraded by financial analysts and losing shareholders. It should be noted that these shareholders can also, like customers, be attracted to the ethics advocated by a company. At the same time, the notion of socially responsible investment (SRI), which comes from North America, is a trend that is spreading in economically and socially advanced countries. SRI could ultimately become a significant criterion for choosing shares in mutual funds or pension funds. A number of ethical financial funds such as, in France, AXA-Génération, Crédit Lyonnais's Pactéo offer, and Prado Epargne's Horizon Solidarité already exist.

Pressure from prescribing stakeholders, such as NGOs and unions

The different players involved in social development, such as NGOs and trade unions, seem to have a thorough understanding of the role and power that go with sustainable development. If properly advised, they will undoubtedly enhance their professionalism in terms of marketing and communication. They campaign for the creation of indexes or 'labels' informing the customers and shareholders of companies that respect the requirements of sustainable development and SRI.

For the moment, most indexes are proposed by independent companies, as is the case for company rating indexes, eg Arese Sustainable Performance Index (ASPI), FTSE, which has created four indexes, and Dow Jones Sustainability Index (DJSI). Most of them are designed based on objective criteria relating to the company's compliance with sustainable development and SRI criteria (eg civil society, corporate governance, health customers and suppliers, safety, environment, human resources, and international labour laws).

In addition to labels and criteria, these stakeholders have the power to influence customers, via large-scale communication actions, and lobbying of governments and government agencies. Lobbying can have direct repercussions on the legislation governing the creation of new products. To cite the example of future French pension funds, certain trade unions are already demanding that the companies eligible for these funds meet specific SRI criteria, and respond to their concerns as a social partner, and that these criteria are integrated into the legislation. Corporate marketing teams will be called upon to take counter-lobbying or social communication measures. The exponential increase in the number of social network members only reinforces this concern.

Pressure from public authorities

In France, the Law on New Economic Regulations (LNRE) is shaking up certain domains. The chapter on the organization and structure of the boards of directors (BDs) should limit the conflicts of interest and overlapping responsibilities and help eradicate the image of BDs as a mere rubber stamp, an issue raised during the Vivendi case.

Another domain tackled by the LNRE is the information that all listed companies must provide in their social and environmental report. The implementing decree of the LNRE provides detail on this information, which includes:

- figures on recruitment, redundancies and professional mobility;
- organization of working time;
- figures on working hours, absenteeism and training;
- details of occupational risks and prevention policy;
- remuneration and types of contract;
- integration of disabled people and gender equality;

- figures on air, water and soil discharge;
- nuisances;
- environmental management organization.

At global level, the United Nations (UN) is reflecting on the standard information that companies should provide as part of the Global Reporting Initiative (GRI) project.

Annual reports tend to include an increasing number of indicators for the benefit of shareholders. All these indicators and this information are provided on the basis of self-assessment, which can be challenged. However, even though companies are not legally required to embrace sustainable development, featuring environmental and social indicators will clearly motivate them to make an effort so that these indicators reflect the reality.

An increasing number of laws relative to sustainable development are emerging. These laws concern in particular the environment and social and economic domains. Europe is increasingly concerned about this issue: today's White Paper will be tomorrow's directives.

The objective of Neuromarketing is to help companies develop pertinent policies in light of the important issues of sustainable development and SRI. It already seems necessary to embark on a reflection in the form of multidisciplinary project teams so as to anticipate problems and prepare pertinent responses to the various hypotheses and alternatives that may arise. In particular, Neuromarketing must examine several subjects such as:

- the relevance of sustainable development and SRI to the customers' conscience and the evaluation of their level of support;
- the repercussions of the damage caused by stress on the company's image, in particular when it leads to an abnormal number of suicides within the company;
- the prescribing power and influence of stakeholders like NGOs or trade unions;
- the foreseeable evolution of national, European and international regulations;

- the influence on product and service policies;
- the influence on the recommendations of financial analysts and shareholders' choice;
- the need to implement a lobbying and institutional communication policy as well as a products and services policy;
- the implementation of internal organization capable of responding to SD and SRI requirements, similar to that adopted for the ISO quality certification, and the possibility of creating an ISO 14004 integrating the response to SD and SRI issues.

Relevance of sustainable development to the company

There is good reason why a significant number of leading companies in Europe are taking an interest in sustainable development. The Johannesburg summit was marked by the strong desire of some of them to commit to this cause.

As noted by Balmadier (2003), 'sustainable development begins as a company project and evolves into a company philosophy'. The companies' interest in sustainable development deals with several concerns:

- *It constitutes an ideal way to create internal cohesion and mobilize the personnel and networks based on major ethical causes.* These companies can focus on the search for cohesion within a national or international context. For companies such as AXA and HSBC that are trying to adopt common values on the group's global stage, this is an opportunity to use the sustainable development cause as a unifying tool.

- *It contributes to improving internal productivity.* Environmental protection helps reduce energy consumption. The obligation to recycle waste helps avoid waste. According to a study by the French Ministry of Employment, good social management is seen by financial analysts as a profitability factor. The search for new processes to adapt to SD imperatives helps develop a spirit of creativity and innovation within the company.

- *It protects against the hazards emanating from major risks.* Respecting financial ethics and acquiring efficient tools to combat money laundering can protect companies against litigation and the deterioration of their image. Refusing customers, suppliers, distributors and intermediaries who do not respect SD and SRI can protect companies against accusations of involvement in misguided causes, resulting in the consequences mentioned above.

- *It helps improve the profitability of products and services as well as customers, by reducing risks, in particular industrial risks.* Companies respectful of sustainable development are less exposed to certain major risks. A concern for continuous economic, environmental and social improvement can only contribute, because of collateral effects, to reducing industrial and also financial risks by improving the overall management of the company.

- *It is a source of innovation and creation for the proposition of new products and services.* Increased interest in the requirements emanating from sustainable development inevitably leads to a concern for prevention. A substantial risk anticipation and risk management market will constitute a source of reflection for the Neuromarketing of tomorrow.

- *It results in enhancing the loyalty of certain customers and suppliers and provides an original and pertinent sales argument for companies strongly involved in this domain.* This sales argument can also be used in the event of mergers or collaborations of mutual insurance companies advocating common SD and SRI values. It attracts a growing number of customers made aware of SD and SRI causes. By agreeing to become the partners of companies that subscribe to these causes, they leave their brain with a clear conscience.

Beyond sustainable development: human rights marketing

The company's legitimacy can be rooted in its profound adherence to the imperatives of sustainable development, social responsibility vis-à-vis its employees and socially responsible investments. Laurent Maruani, Professor of Marketing at HEC Paris, believes that, beyond this theme, the marketing of tomorrow must strongly contribute to the defence of human rights. He outlines his recommendations in an interesting article published in *La Tribune* (Maruani, 2009).

It is possible, using a traceability indicator, to reconcile the market approach and the moral approach by highlighting the power of consumers. This is of great benefit to socially responsible firms, as they avoid direct confrontation with partners and countries that do not respect human rights, resulting in their customers' lack of interest in their offers for moral or ethical reasons. For the customers, the company's legitimacy is reinforced, as it leaves their intelligence with a clear conscience when they purchase. For the employees, it reduces the level of stress in the often expatriated personnel concerned, who are sometimes obliged to propose products manufactured in a way that conflicts with their personal ethics.

The proposition consists of cautiously and rigorously introducing a 'human rights' traceability indicator and featuring it on offers via an independent certification system. With this transparent and objective information, the consumer is free to purchase or not. This marketing approach allows a company to drop an offer that does not respect human rights, not for reasons of morality or its own code of ethics, often powerless in the face of business reality, but because of its own consumers' refusal to purchase the incriminating products. This proposition, advocated by the French ambassador for human rights, is favoured by certain entrepreneurs, whose often expatriated personnel are subject to intolerable pressure. This pressure generates an unnecessary amount of stress, suffering and professional difficulties. Neuromarketing, which strives to create harmony between the marketing approach and the deepest concerns of the intelligence of customers and employees, can only welcome this type of initiative.

PART IV: KEY POINTS

- The marketing approach of the 21st century must adapt to the major developments imposed upon the intelligence of consumers who have become proactive as a result of interactivity.

- This intelligence is subject to multiple changes conditioned by the increasing importance given to emotion over logic, the internet invading our homes, communication messages saturating the customer's brain, the multiplication of offers creating a permanent level of stress in consumers, and citizens' growing interest in corporate social responsibility as well as in the respect of human rights.

- To address these developments, the marketing approach has to evolve and focus its efforts on new domains. Value innovation allows marketing to surprise the customer's brain by offering new horizons via the companies' offers. When they try to be not just better but different, the marketing propositions become exceptional, because they resemble no other services provided by the competition. By aiming for disruption or the search for the 'blue ocean', marketing is reinventing traditional strategic approaches and encouraging companies to improve their creativity considerably. This makes it possible to provide the brains of executive committee members with an innovative business model, and to provide the customer's brain with a unique positioning likely to stimulate the customer's desire for the offer proposed.

- Permission and desire marketing is developed to avoid saturating the brain with repeated, intrusive and unwanted offers and messages. To appeal to customers' intelligence, it does more than simply target their needs: it takes an interest in their desires. This new marketing approach refuses to communicate with customers who have not given their permission. By doing so, it prevents companies from spending considerable budgets that will end in spam or unwanted messages, which only result in irritating or exasperating the customer's intelligence.

- Permission and desire marketing help find and renew highly desirable bait for the customers, making it possible to get their consent to communicate with them. It proposes a method that, by way of the internet, helps reach a maximum number of community prescribers and uses viral marketing or 'buzz marketing'.

- Interactive marketing paves the way for conversation marketing by facilitating dialogue with the customer's brain. It is largely based on the use of e-marketing on the internet. It is made possible by the use of customer relationship management (CRM). It supports the dynamic management of powerful databases. It challenges the principles of large-scale traditional segmentation and the marketing mix by attempting to initiate mass customization or one-to-one marketing.

- Brand marketing, for its part, is designed to reassure the customer's brain. It strives to enhance customers' loyalty and get them to recommend the brand to acquaintances, friends and the internet community to which they belong. More than just communication, it helps define, for the company, its offers and sales platforms, and a genuine brand policy. It is based on the strategic triad of positioning, identity and image. It proposes a marketing mix of the brand as well as a suitable organization based on a significant change in mentalities.

- The final contribution of 21st-century marketing presented in Part IV is a focus on the quality and legitimacy that the company must acquire for the intelligence of responsible customers. This reassures the brain, which has given its approval for the purchase, while leaving its conscience clear to embark on an emotional relationship with a company, product or service brand. Its purpose is to help retailers adapt the resources that will legitimize them in the eyes of consumers, represented by their adherence to the principles of sustainable development, fair trade, social and societal responsibility, and human rights.

Vision of
the future

Human intelligence does not really envisage the future. Its attitude towards tomorrow's innovations is more often doubtful and sceptical than confident and hopeful.

In the early 20th century, if anyone had announced that people would fly and go to the Moon, cure certain diseases that were incurable at the time, or watch things happening in the world on a colour screen sitting at home in an armchair, they would have been suspected of acute dementia. Closer to our time, in the 1970s, a visionary announcing the standardization of the mobile phone and the internet would probably have been considered mad.

These unimaginable innovations did occur. With the acceleration of progress, the impossible is happening at a blistering pace. Progress is now so fast that, contrary to ancient habits, children are educating their parents. This is the case with the use of the internet, smartphones and social networks.

To be open to the future, modern intelligence must integrate, as was the case in the early 20th century, the significant divide that will irrevocably grow between the capabilities of the imagination and the advent of progress.

Based on the wealth of research currently conducted in laboratories throughout the world, certain sciences may profoundly disrupt our knowledge and lifestyle in the near future:

- *Physics*, with a better knowledge of the infinitely small and infinitely large, which tend to converge.

- *Computer science*, with the possibility of storing and processing an almost limitless amount of information in increasingly confined and affordable spaces. It is currently possible to store as much information in a Sony PlayStation console worth a hundred euros or so as in a computer room in the 1970s that would have cost several million dollars.

- *Robotics*, with the design of machines that resemble human behaviour more and more every day.

- *Bionics*, which strives to bring the computer closer to the human brain. Some researchers believe that, in less than 30 years, it will be possible to connect the brain directly to a computer, using a very small and embedded communication chip controlled by thoughts.

- *Genetics*, which enhances knowledge of the cell's components and DNA. It is speculated that human memory is passed on between generations by genes received at birth. These assumptions would explain dreams about unknown references and could challenge certain philosophies and beliefs. Metempsychosis, the reincarnation of souls, would be explained by the simple transmission of the memory by the genes received from our parents, grandparents, great-grandparents and remoter ancestors.

With the evolution of progress, Neuromarketing, which may seem surprising to some readers of this book, could become commonplace over the next few decades. Medical knowledge of how the brain and hormone secretion work is developing in a spectacular manner. Its application to marketing will rapidly become routine. The objective of the authors when writing this fundamental book on Neuromarketing is not to promote esotericism or become the gurus of a new science. It is, more modestly, to raise the readers' awareness of a new school of thought, new approaches, and reflections that have begun to affect the marketing concept, space and approach in terms of sales and communication.

By creating Neuromarketing, the combination of neuroscience and marketing sheds new light on the studies, the decision-making process and the organization of actions induced by marketing. It provides insight into the reasons behind attitudes already identified or felt, which are often common sense, by marketers, salespeople and communicators. It paves the way for new reflections, which should result in increased research in universities and international schools of management and marketing.

Certain studies are now published in major, internationally renowned scientific managerial and marketing journals. They will undoubtedly intensify in the future. Courses will emerge and multiply. In the words of Victor Hugo, 'When an idea reaches maturity, nobody can stop it.'

Because of its efficacy, Neuromarketing is not without danger: danger of manipulating the customer to great effect. Totalitarian regimes have already understood the benefits they could derive from using the drugs of the brain and do not hesitate to do so. This is particularly true of subliminal forms of communication to render the subject dependent.

Manipulation is all the more effective when some people know techniques unknown by others. Once they become known to the general public, these techniques automatically become less effective. They can even violently turn against those who dare use them. If Neuromarketing becomes manipulative, it is on the road to ruin. As shown in Stage 6 of the Neuromarketing method, Neuromarketing must focus on meeting the customers' actual needs to enhance their loyalty. It must also inform them of the intelligence traps set by sales and communication professionals.

An ethical approach to Neuromarketing is vital for marketers wishing to establish durable, trust-based relationships with their customers, and turn them into genuine partners of their company.

Knowledge of every individual's 'S point' is important for finding, if not wisdom, at least serenity. Tomorrow's research on the brain may teach us that the pursuit of happiness is found in the human intelligence's ability to conceive the life of an individual based on his or her 'S point'.

REFERENCES

Badoc, M and Beauvois-Coladon, M (2008) Les stratégies de l'océan bleu permettront-elles de retrouver les voies de la croissance? [Will blue ocean strategies help us restore growth?], *Revue Banque*, 707

Balmadier, R (2003) Les compagnies d'assurance doivent-elles s'intéresser au développement durable? [Should insurance companies be interested in sustainable development?], MBA dissertation, ENASS, Paris

Bard, A and Bard, M (2002) *The Complete Idiot's Guide to Understanding the Brain*, Alpha Books, New York

Bateson, G (1979) *Mind and Nature: A necessary unity*, Advances in Systems Theory, Complexity, and the Human Sciences, Dutton, New York

Changeux, JP and Garey, L (2012) *The Good, the True, and the Beautiful: A neuronal approach*, Yale University Press, New Haven, CT

Chebat, JC and Langlois, M (1989) Le canal bancaire et le service [Channel bank and service], Congrès de l'Association Française du Marketing, 56–64

Clodong, O and Chétochine, G (2010) *Le storytelling en action* [Storytelling in action], Eyrolles, Editions d'Organisation, Paris

Collet, D, Lansier, P and Ollivier, D (1991) *Objectif zero défaut: Mesure et qualité totale dans le tertiaire* (Zero defect objective: Measurement and total quality in the tertiary sector), Editions ESF, Paris

Damasio, A (1994) *Descartes' Error: Emotion, reason and the human brain*, Avon Books, New York

Damasio, A (2005) *Descartes' Error*, Penguin, New York

D'Aveni, RA (1994) *Hyper-Competition: Managing the dynamics of strategic maneuvering*, Free Press, New York

De Bono, R (1996) *Serious Creativity*, HarperCollins, New York

Dietvorst, RC *et al* (2009) A sale force-specific theory-of-mind scale: Tests of its validity by classical methods and functional magnetic resonance imaging, *Journal of Marketing Research*, **46** (5), pp 653–68

Dooley, R (2012) *Brainfluence: 100 ways to persuade and convince consumers with neuromarketing*, John Wiley & Sons, Hoboken, NJ

Dru, JM (1997) *Disruption: Overturning conventions and shaking up the marketplace*, John Wiley & Sons, New York

Dufour, A (1997) *Le cybermarketing: Intégrer internet dans la stratégie d'entreprise*, Que sais-je, no 3186, Presses universitaires de France, Paris

du Plessis, E (2011) *The Branded Mind: What neuroscience really tells us about the puzzle of the brain and the brand*, Kogan Page, London

Eagleman, D (2011) *Incognito: The secret lives of the brain*, Canongate Books, Edinburgh

Gazzaniga, M (2008) *Cognitive Neuroscience: The biology of the mind*, WW Norton & Co, New York

Georges, P (2004a) *Etre au top*, Eyrolles, Paris

Georges, P (2004b) *Gagner en efficacité* [Win in efficiency], Editions d'Organisation, Paris

Georges, P (2005) *Vaincre le stress*, Eyrolles, Paris

Giboreau, A and Body, L (2007) *Le marketing sensoriel* [Sensory marketing], Vuibert, Paris

Ginger, S (2007) *Gestalt Therapy: The art of contact*, Karnac Books, London

Godin, S (1999) *Permission Marketing*, Simon & Schuster, New York

Godin, S (2002) *Purple Cow: Transform your business by being remarkable*, Penguin Group, New York

Godin, S (2005) *All Marketers Are Liars: The power of telling authentic stories in a low-trust world*, Penguin Group, New York

Handy, C (2002) *The Age of Unreason: New thinking for a new world*, Arrow, London

Hedgcock, W and Rao, AR (2009) Trade-off aversion as an explanation for the attraction effect: A functional magnetic resonance imaging study, *Journal of Marketing Research*, **46** (1), pp 1–13

Hultén, B, Broweus, N and van Dijk, M (2009) *Sensory Marketing*, Palgrave Macmillan, Houndmills

Kahneman, D (2011) *Thinking, Fast and Slow*, Farrar, Straus and Giroux, New York

Kapferer, J-N (1991) *Les marques: Capital de l'entreprise* [Brands: Equity of the firm], Editions d'Organisation, Paris

Kapferer, J-N (1994) *Strategic Brand Management: New approaches to creating and evaluating brand equity*, Free Press, New York

Kapferer, J-N (1998) *Les marques: Capital de l'entreprise* [Brands: Equity of the firm], Editions d'Organisation, Paris

Kapferer, J-N (2008) *The New Strategic Brand Management: Creating and sustaining brand equity long term*, 4th edn, Kogan Page, London

Kim, W Chan and Mauborgne, R (2005) *Blue Ocean Strategy: How to create uncontested market space and make the competition irrelevant*, Harvard Business School Publishing, Boston, MA

Krishna, A (2009) *Sensory Marketing: Research on the sensuality of products*, Routledge Academic, New York

Le Bihan, D (2012) *Le Cerveau de Gustal: Le que nous révèle la neuro-imagerie*, Odile Jacob, Paris

Lecerf-Thomas, B (2009) *Neurosciences et management* [Neurosciences and management], Eyrolles, Paris

Lee, Paul and Stewart, D (2012) *Technology, Media and Telecommunications Predictions 2012*, Deloitte Global Services, London

Le Guérer, A (2005) *Le parfum: Des origines à nos jours* [Perfume: The origins to the present day], Odile Jacob, Paris

Le Nagard-Assayag, E and Manceau, D (2011) *Le Marketing de l'innovation* [Innovation marketing], Dunod, Paris

Levine, R *et al* (2009) *The Cluetrain Manifesto*, Basic Books, New York

Lewi, G (2003) *Les marques: Mythologies du quotidien* [Brands: Daily mythologies], Village Mondial, Paris

Lindstrom, M (2010) *Buyology: Truth and lies about why we buy*, Broadway Books, New York

MacLean, P (1990) *The Triune Brain in Evolution: Role in paleocerebral functions*, Plenum Press, New York

McDonald, MHB and Morris, P (1992) *The Marketing Plan: A pictorial guide for managers*, Heinemann Professional Publishing, London

McLure, SM *et al* (2004) Neural correlates of behavioral preference for culturally familiar drinks, *Neuron*, **44**, pp 379–87

Maruani, L (2009) Droits de l'homme et marketing peuvent faire bon ménage, *La tribune*, Supplement to no 4297, 25 September

Moiroud, R (1993) *Le cri du client* [The customer's shout], Editions d'Organisation, Paris

Montague, R (2007) *Your Brain Is (Almost) Perfect: How we make decisions*, Plume, New York

Morris, D (1999) *The Naked Ape: A zoologist's study of the human animal*, Dell Publishing, New York

Ornstein, R (1992) *Evolution of Consciousness: The origins of the way we think*, Simon & Schuster, New York

Peppers, D and Rogers, M (1996) *The One-to-One Future*, Currency Doubleday, New York

Peters, T (1987) *Thriving on Chaos: Handbook for a management revolution*, HarperCollins, New York

Peters, T and Waterman, R (1982) *In Search of Excellence*, Harper & Row, New York

Plassmann, H, Ramsøy, TZ and Milosavljevic, M (2012) Branding the brain: A critical review and outlook, *Journal of Consumer Psychology*, **22**, pp 18–36

Pradeep, AK (2010) *The Buying Brain: Secrets for selling to the subconscious mind*, John Wiley & Sons, Hoboken, NJ

Proust, M (2003) *In Search of Lost Time*, Proust Complete, Modern Library, New York

Rancev, C (2002) *Qvels sont les Enjeux de la marque?* Master thesis in Bank, Finance and Insurance, Paris X Nanterre University

Rava-Reny, F (2003) Le cerveau triunique de MacLean [MacLean's triunic brain], *Intelligence mode d'emploi*, 7.

Raveleau, G (1990) *Journal des Caisses d'Epargne*, Special issue

Rechenmann, JJ (2001) *Internet et marketing* [The internet and marketing], Editions d'Organisation, Paris

Reichheld, F (1996) *The Loyalty Effect*, Harvard Business School Press, Boston, MA

Renvoisé, P and Morin, C (2002) *Selling to the Old Brain*, SalesBrain, San Francisco, CA

Rieunier, S *et al* (2009) *Le marketing sensoriel du point de vente* [Sensory marketing in the retail industry], 3rd edn, LSA, Dunod, Paris

Rizzolatti, G, Sinigaglia, C and Andersson, F (2008) *Mirrors in the Brain: How our minds share actions, emotions and experience*, Oxford University Press, New York

Rosen, E (2000) *The Anatomy of Buzz: How to create word of mouth marketing*, Doubleday, New York

Rosen, E (2009) *The Anatomy of Buzz Revisited*, Doubleday, New York

Saint-Exupéry, A de (2000) *The Little Prince*, Mariner Books, New York

Séguéla, J (1982) *Hollywood lave plus blanc* [Hollywood washes whiter], Flammarion, Paris

Servan-Schreiber, D (2012) *Healing without Freud or Prozac: Natural approaches to curing stress, anxiety and depression*, Pan Macmillan, London

Singer, S (2007) *Gestalt Therapy: The art of contact*, Karnac Books, London

Steidl, P (2012) *Neurobranding*, CreateSpace Independent Publishing Platform, Lexington, KY

Thomas, RJ (1995) *New Product Success Stories: Lessons from leading innovators*, John Wiley & Sons, Toronto

Van Aal, J (1981) *Connivence: Une autre façon d'être publicitaire* [Connivance: Another way to be an advertiser], Luneau Ascot Editeurs, Paris

Variot, JF (1982) *Huit stratégies publicitaires: Pour gagner face à la crise* [Eight advertising strategies: How to win facing the crisis], Editions d'Organisation, Paris

Variot, JF (2001) *La marque post publicitaire: Internet, acte II* [The post-advertising brand: The internet, Act II], Village Mondial, Paris

Walder, F (1959) *The Negotiators*, McDowell, Obolensky, New York

Weinschenk, SM (2009) *Neuro Web Design: What makes them click?*, New Riders, Berkeley, CA

Zurawicki, L (2010) *Neuromarketing: Exploring the brain of the consumer*, Springer, New York

INDEX

Lightning Source UK Ltd.
Milton Keynes UK
UKHW011914230819
348474UK00004B/150/P